SOUTHERN · TIMES ·

Contents

The next issue of Southern Times, No 14, will contain:
Gosport Part 2, Stephen Townroe; Whitchurch 1954,
Southern 4-6-0's, some LSWR wagons,
the last years at Ilfracombe in colour, Ocean Liner trains,
more from the Southern Railways Archive,
Brighton line Part 2, From the Footplate etc etc.

The Transport Treasury

TIMES SERIES

Front Cover: 'The Yanks are here!' Three members of the USA class waiting their next duties in the Western Docks at Southampton on 26 June 1957; Nos 30070, 30063 and 30072. No 30063 was withdrawn due to collision damage and scrapped but the other found their way into preservation. *R C Riley / Transport Treasury*

Above: Forecourt views are not something we have featured for a little time so it is with pleasure that we include this view of Marden station on the South Eastern main line and serving the area of the same name near to Maidstone. Opened in 184? it remains in use today but with basic up and down main lines and no trace of the goods facilities seen here. Indeed the goods yard is extremely neat and tidy - note the coal sacks over the end of the buffer stop. At this point in history, c1950s, the location still had the typical SER staggered platforms although these have been replaced today. The coal wagons may well have been unloaded and are waiting collection - no demurrage payments here! Station Masters were taught to press home the need to unload wagons in a timely manner sometimes using the phrase, 'If it aint turning it aint earning'. *Transport Treasury*

Rear Cover: Former Wainwright tender, renumbered as DS70199 and re-purposed for drinking water - we would hope boiled before use! No details as to where or when but Ashford appears to be the most likely choice and sometime in the mid 1960s. (Further information welcome.) *Terry Tracey / Transport Treasury*

Copies of many of the images within **SOUTHERN TIMES** are available for purchase/download.

In addition the Transport Treasury Archive contains tens of thousands of other UK, Irish and some European railway photographs.

© Kevin Robertson. Images (unless credited otherwise) and design The Transport Treasury 2025

ISBN 978-1-917776-11-0

First Published in 2026 by Transport Treasury Publishing Ltd., 16 Highworth Close, High Wycombe HP13 7PJ

www.ttpublishing.co.uk or for editorial issues and contributions email to **southerntimes@email.com**

Printed in the Short Run Press, Exeter.

INTRODUCTION

I have mentioned previously that compiling 'Southern Times' (and its predecessor 'Southern Way') has been a voyage of discovery.

With any interest it can be all too common to think one knows it all, or certainly enough to satisfy the demand but the fact you are reading this likely means you are like me with an open mind for subjects that had perhaps not been considered.

This is where the contributions to ST are so readily received from those who have specific knowledge on their local area regardless of period and who are prepared to share it with others. As an example I fully admit I will never have the in depth knowledge of say a reader from the Dover / Folkstone area (one of many locations that I struggle on) but we try and I hope in every issue there is perhaps at least one nugget to make the reader sit up and say to himself, 'I never knew that'.

To compile ST then, we need new information and I can promise we are currently negotiating for something very special from the Chief Civil Engineer's Department. Harking back to this aspect of the railway, I recall in the 1980s spending hours in the plan arch at Waterloo where dear Reg Randall and his able assistant Derek Clayton would patiently answer questions and produce plans from the drawers of chests still having labels from the pre-group companies. Both of them had an encyclopedic knowledge of the Southern from east to west and north to south and never were they defeated with a 'where is it question' when shown a photograph. Numerous books and articles owe their origins to these men each of whom has now passed. A few weeks later I am delighted to report our negotiations reached a satisfactory comclusion and we are now the (temporary) custodians of a large number of illustarted files depicting bridges mainly on the former South Western lines in Surrey, Hampshire and Dorset. A sample selection will appear as the lead article in Issue 14.

I am sometimes asked, 'Where do you get your information from?' and the answer is numerous sources. Sometimes it is luck as was indeed the case when Iain Steele of the 'Friends of Swindon Museum' passed me an A4 file of Southern papers; not something Swindon would retain. Something about the contents made me take a literal step back as here in fact was an unpublished manuscript and one from someone who was very special with a reputation second to none - D L Bradley.

I doubt there is anyone reading this who has not heard of Bradley, author of locomotive books on the constituent companies of the Southern as well as two books on Southern locomotives; Maunsell and Bulleid. Over the years some of these have been enlarged and reprinted (the LSWR books by Wild Swan and more recently his 'Locomotives of the Isle of Wight Railways'. Irwell Press especially have also taken specific classes as the subjects for individual books all having their origins in the work of Bradley. Indeed I am sure I am not alone in often going back to Bradley as the first reference on locomotive matters.

There was however one area of Southern Locomotive history that Bradley did not cover. This was Bulleid's designs for Leader, the main line electric locos, main line diesels and his one-off diesel shunter. As to why there was nothing was the subject of rumour and speculation but that can now be countered as we have *FOUND* Bradley's missing manuscript on these engines. It makes for fascinating reading as well and whilst the first three named designs have been the subject of separate books the way Bradley describes each is still worthy of study. Consequently I am delighted to say we will be printing it in the same style and format (with the knowledge and consent of the RCTS) to be released in 2026.

As it stands it is a concise work, something over 45,000 words. There are no illustrations just some photocopies of suggested views; we can certainly find images with the intention to produce the work to the same size and style as the two earlier 'Locomotives of the Southern Railway' books.

At this stage I can imagine the question, 'Where on earth did you find this..?' Well as I mentioned Iain Steele located it at a house clearance in the Swindon area, the previous owner a bit of hoarder who would literally go to boot sales and the like to 'hoover up' anything remotely railway related. As to why it was never published we have no idea and enquiries in that area have drawn a blank. Fifty years after publication of the first to book we feel we can at last complete the series.

Kevin Robertson

Top: At Eastleigh, the Down 'Bournemouth Belle' is headed by Bulleid/ English diesel locomotive No 10203 The start cf t~e Gosport branch is on the right by the coaches in the carriage sidings. The two lines on the right are the access / exit roa:s from the running shed. The Works entrance and exit just by the trees on the right. *R E Vincent / Transport Treasury*

Bottom and opposite: Botley station in 1968. There was once heavy seasonal strawberry traffic from this bay. The narow Down platform held the Bishops Waltham branch train in the bay, for many years a rail-motor. It ran behind the signal box, curv ng right by the far road bridge (opposite). Today the station has lost its brick building and timer shelter in favour of 'bus' type she ters. *Both: John Scrace / Bluebell Railway Archives*

Pilgrimage to Gosport - Part 1 (North)
Alan Postlethwaite

The London and Southampton Railway opened to Eastleigh in 1839 and to Southampton in 1840. The Gosport branch opened in 1841, becoming the first railway to serve Portsmouth, albeit by ferry. The LBSCR reached Portsmouth Town in 1847 and the LSWR followed the following year. The branches to Bishops Waltham and Stokes Bay opened in 1863, the latter offering the shortest crossing to the Isle of Wight. The Fareham to Southampton line opened in 1889 and the branch to Lee-on-the-Solent in 1894. Finally, the Meon Valley line opened in 1903, offering an alternative route to and from London.

Bored through clay, Tapnage tunnel, commonly known as Fareham tunnel, encountered major problems of slippage and collapse both during construction and in operation. It was relined in the 1870s, making it narrower. Even so a deviation line was built, half a mile west of the tunnel. The tunnel was singled in 1906 to be used exclusively by Meon Valley trains. The deviation line also suffered from slippage and was closed in 1974.

Fareham was a busy junction station, used by local trains and by through traffic from Portsmouth, Brighton and Hastings to the West of England and South Wales via Southampton or Eastleigh. Its long goods yard included a gantry crane.

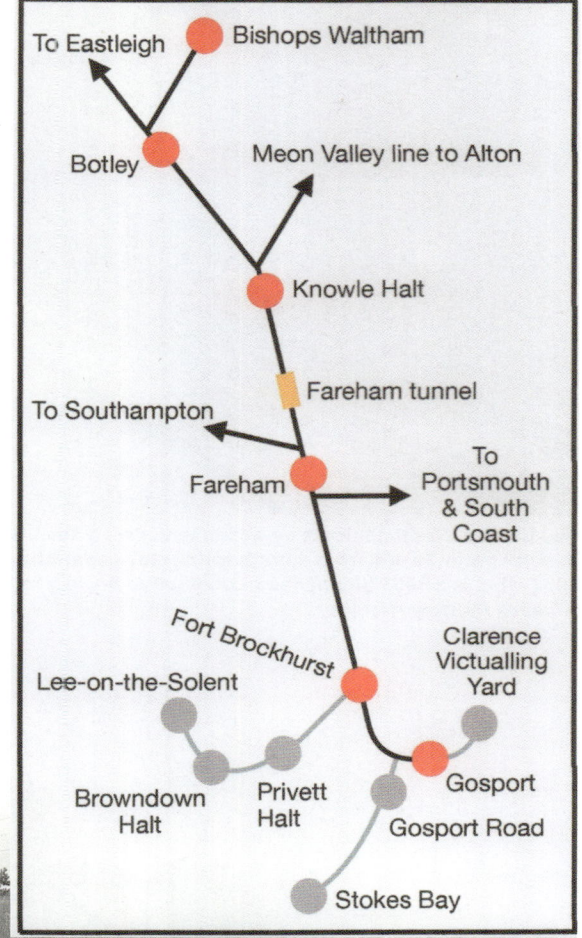

The Bishops Waltham Railway was taken over by the LSWR in 1881. Passenger traffic was light but the line was kept busy with mineral traffic to and from a brickworks, clay pits and a gasworks. The branch closed to passengers in 1933 and to all traffic in 1962. This is a 1953 Stephenson Locomotive Society rail tour between Havant, Gosport and Bishops Waltham. *John J Smith, Bluebell Railway Archives*

Bishop Waltham's compact but substantial station buildings have the most beautiful of chimney stacks.This was a proud branch line with the terminus recorded in 1950. (The stretcher and first aid cupboard still stand under the awning.) The view is looking towards the goods shed and the end of the branch. *Denis Cullum Collection, Lens of Sutton Association*

T9 No 30283 on a local service near Botley bound for Fareham in 1957 (the headcode alos used to apply to Gosport trains). *Colin Hogg, Bluebell Railway Archives*

S15 No 30837 heads a 1966 LCGB rail tour on a circular tour from Waterloo via Bordon, Alton, Eastleigh Works, Fareham, Havant and back to Waterloo. *John Scrace, Bluebell Railway Archives*

The solitary preserved N15 No 30477 *Sir Lamiel* on a van train near Botley in 1957. *Colin Hogg, Bluebell Railway Archives*

Knowle Halt served a psychiatric hospital. Opened in 1907 and closed in 1964, it was located by the Meon Valley track comprising a simple platform with a bench and shelter. The long Down freight is headed by class N15 No 30451, *Sir Lamorak*.
Lens of Sutton Association, 1955

Looking north from Fareham's island platform, these starting signals are for the Southampton line (L), the tunnel deviation line (centre) and the tunnel line (R). *Alan Postlethwaite, 1958*

LSWR class 700 No 30308 prepares to depart Fareham with the 6.48 pm service to Alton sometime in 1954. On the left is the goods shed. This was the last full year of operation on the Meon Valley line. *Colin Hogg, Bluebell Railway Archives, 1954*

Billington's class K was arguably the most handsome Mogul on the South Coast. No 32337 heads a heavy Down freight through Fareham. Note the theatre-type route indicator on the right which could display the letter 'A' for Alton bound trains. *Alan Postlethwaite, 1958*

Diverted due to engineering works south of Eastleigh, Battle of Britain class No 34082, *615 Squadron*, arrives at Fareham with a Bounremouth express from Waterloo having arrived via Botley. West Country No 34097, *Templecombe*, will then atach to the opposite end to take the train on to Southampton and Bournemouth via Netley and on to Southampton and Bournemouth.
Joe Kent, Bluebell Railway Archives, 1962

At Fareham, Drummond D15 No 30471 heads the 12.23 pm from Southampton to Portsmouth, a contender perhaps for the grimiest train in the UK! *Transport Treasury*

Pilgrimage to Gosport - Part 2 (South)
by Alan Postlethwaite
will appear in the next issue.

In the interim here is a reminder of Gosport from the 1950s. T9 No 30175 crossing Spring Garden Lane at Gosport en-route to the admiralty yard on 27 June 1950. The station's overall roof was meant to be a temporary replacement to the original which was destroyed, along with much of the station, in an enemy raid overnight on 10/11 March 1941, (the roof was never made permanent). The incumbent Station Master also lost all his furniture in the ensuing blaze; he had been due to move the very next day with all his goods and chattels already packed in a railway container. *R C Riley / Transport Treasury*

Special train at Gosport on 7 March 1959 organised by the Branch Line Society. Regular passenger servcies had already ceased six years before in 1953. Starting at Portsmouth Harbour, the trip took in the stub of the Midhurst line from Chichester as far north as Lavant, from Botley to Bishops Waltham, Fareham to Droxford and which was as far north on the Meon Valley line as it was possible to travel, and Fareham to Gosport before returning to Portsmouth Harbour. M7 No 30111 was the motive power together with pull-push set No 6 (coach Nos 6496 and 1103). *A E Bennett / Transport Treasury*

Gratuitous plugs next, but previously published titles relevant to the subject in hand include:

'The Bishops Waltham Branch'. Roger Simmonds and Kevin Robertson. Wild Swan Publications. 1988.

'The Meon Valley Railway Revisited'. Denis Tillman. KRB Publications. 2003.

'The Meon Valley Railway'. Ray Stone. Kingfisher Publications. 1996 (reprint).

'The Meon Valley Railway Part 1 Building the Line'. Kevin Robertson. Noodle Books 2011.

'The Meon Valley Railway Part 2 A Rural Backwater'. Kevin Robertson. Noodle Books 2012.

'The Meon Valley Railway Part 3 Closure and Beyond'. Kevin Robertson. Noodle Books 2013.

'The Railways of Gosport including the Stokes Bay and Lee-on-the-Solent Branches'. Kevin Robertson.

Kingfisher Publications 1986 reprinted 2009

The subject that just will not lie down: No 36001
Further thoughts by Guy Cooper

Traditionally it is in the introduction to a magazine or journal where the editor is supposed to explain or excuse his or her behaviour. On this occasion I felt it more appropriate to do so here, for yet again we turn to, as the heading states, 'The subject that just will not lie down', No 36001 or 'Leader'.

I will also fully admit this is likely entirely my fault. Agreed it is a pet subject but also one which seems to continue to attract interest from others. Witness then Francis Terry in ST4 with new (and proven) evidence of track-spreading between Brighton and Lewes. In issue 9 with the article on the (nightmare) building programme of 1946 accompanied by the thoughts of retired engineer John Wenyon. Then in ST12 we had a most interesting opinion by Jim Gosden contained within 'From the Footplate' and which prompted reader Guy Cooper, himself another retired engineer to put forth his own ideas.

Having read Guy's text I was impressed. In short his opinion is very much in tune with my own. Guy was concerned that in some respects it may be too controversial, too apparently critical but I hope I have assuaged him to refrain from alteration.

I am delighted to present an unabridged version.

The LEADER – Oh what a controversial locomotive!

Some 80 years on, we can look back at the Leader project as history. We have the benefit of hindsight which we can apply with wisdom, not to criticise but try to analyse and to some extent speculate on what went awry.

Firstly, let's be clear, Oliver Bulleid was certainly a brilliant engineer. Witness his: (a) Q1 0-6-0, a most effective development of Maunsell's Q class. Interesting as their appearance may be, they did their job 'out of the box' and until the end of steam. (b) The electric locos 20001-3 worked well, doing the work for which they were intended. (c) The big diesels 10201/2 were similarly a successful prototype, a forerunner of what was to come, with features being copied for early production BR diesels. (d) All his loco boilers were great steam raisers. (e) And his adoption of boiler feed water treatment was very effective to improve the availability of steam locomotives. There are many other things that show OVB was a very capable engineer.

Now we get to the Leader Project.

The Design Evolution.

In 1944 the Southern Railway issued a requirement, as E.S. Cox writes "Towards the end of the war the Southern Railway was considering a new type of tank engine, ... in our own records, for the purpose of replacing the M7 Class 0-4-4 Tanks" (ref 1, p. 17).

As I am sure readers know, the Drummond M7 0-4-4T tank engine was introduced in 1897, becoming BR class 2, thus of moderate power. The class was limited from higher speeds after the 1898 Tavistock derailment (ref 2, p.163) and found very suitable for branch line passenger work and also adopted for empty carriage movements. What would Oliver Bulleid design for its replacement?

A sensible first consideration for a new loco is to decide if there is something successful in existence that is already adequate, or that can easily be adapted to fulfil the new requirement, and this is exactly what Mr Bulleid did. An early proposal was to modify the tender of the Q1 for better rearward visibility which would admirably have suited empty carriage stock movements. But back to the need for a new branch-line loco.

Oliver Bulleid's initial thinking was for a tank engine version (ref 3, p. 27) of his successful Q1. But since the Q1 is a class 5 loco, is it a bit too powerful in the need to replace a class 2 M7 on branch line duties? Already we see the design proposals drifting away from the specified requirement.

Seen from ground level, No 36001 presents an impressive some might say formidable presence. Hardly the look of any previous steam engine some details stand out and are worth mentioning: the shade for light above the coupling (just under the cast number plate) - a useful detail at night, the curved access steps - not such a good idea for whilst certainly necessary for access / egress if the engine were on a curve, how many footplate-men caught their shins on them is not reported. Note also that whilst the sleeve oscillating gear has been removed someone has taken the trouble to maintain the carriage steam-heat hose connection. For interest sake, the position of the whistles was not symetrical at both ends. *Pursey Short*

Just how did an M7 morph into 36001...?

Further developments of the Q1 tank engine concepts arrived at a 4-6-4 tank engine (ref 3, p. 20 & 21). No doubt the 4 wheel bogies resulted from George Ellson, the Civil Engineer, being very cautions since the 2-6-4T River Class fatal crash of 1927. But a 4-6-4T based on the Q1 is far too mighty for branch line use! Colin Boocock identifies a need for outer suburban tank locos (ref 4, p. 113/114) for which the 4-6-4T could be fully appropriate, an example being the heavy commuter trains on the quite hilly Oxted line. The design is now drifting well away from the need to replace M7s.

Let's now consider the wheel arrangement suitable for branch use. The 0-4-4T wheel arrangement was quite satisfactory for all existing M7 duties, so, by what in the aircraft industry is called grandfather rights, Mr Bulleid could have repeated the 0-4-4T. However repositioning an axle to 2-4-2T would no doubt have pleased the Civil Engineer, or similarly extending to a small 2-6-2T would have also increased the proportion of adhesive weight. The GWR was then very satisfactorily using 1400 class 0-4-2T locos for branch lines. There is no record of Mr Bulleid contemplating a small loco for the duty.

Frustrated with the low proportion of the 4-6-4 loco weight being available for adhesion, Oliver Bulleid allowed his fertile mind to go into into hyper-drive producing the total adhesion 0-4-4-0 (ref 3, p. 21), and when his draughtsmen were unable to keep below the 20 ton axle load, the 0-6-6-0 Leader was conceived. Mr Bulleid's justification for this new design was to satisfy 10 very laudable objectives, one of which was for a loco "To be capable of working all classes of trains up to 90 mph" (ref 5, p. 217). This had to be a loco of similar capability to the company's existing big 4-6-0 and 4-6-2 tender engines, and as such the Leader would weigh around 120 tons. Nothing could be further from an M7 replacement! Bulleid knew it. That is why he went to the directors to obtain approval for his new and completely different proposal, the Leader.

The SR Directors Response to Leader.

Despite being the exact opposite of the directors' M7 replacement specification, Bulleid was clearly convinced that this is what the company needed. But why, oh why, did the directors agree? What a clever political ploy by Bulleid, the directors were now complicit in the Leader project. Had the directors not realised that it was hopelessly optimistic for Bulleid to think that he could successfully fulfil these idealistic objectives in a single design?

Why did the directors not say to Bulleid "No! – Our request was for an M7 replacement, please provide what we still want"? Then perhaps follow with "we can separately evaluate the benefits and risks of your new big engine proposal"? The directors must have been bright enough to realise that Bulleid's proposal was completely unsuitable for M7 duties. It is an unsolved mystery why they agreed to the Leader. To this writer, that the directors approved the build of 5 Leader locos, does suggest that they thought that the new engine was a production proposition rather than a highly experimental machine of many untried novel features. Bulleid sold the Leader concept advantages to the directors. Had he not also advised them of the risks involved in the very adventurous design concepts going into Leader? Pretty unlikely.

Let's consider Bulleid's position.

Here are 3 previous experimental locos he must have known about, he even worked on the W1:

1. M Rly Paget 2-6-2 No. 2299 (1908). Also featured sleeve valves and a fire brick lined firebox.

2. LMS 4-6-0 Fury No. 6399 (1929). Very high pressure boiler.

3. LNER 4-6-4 W1 No. 10000 (1929). Yarrow high pressure boiler.

None of the novel features of these locos achieved the hoped for benefits, and the former two didn't even enter service. The W1 did, but to continue in service was rebuilt to the remove the novel new features. Bulleid must have been fully aware of the inadequacy of the new concepts on all these experimental locos. Novel features on steam locos were of a high risk of innumerable teething troubles and rarely delivered any of the hoped for benefits. Now, Bulleid's Pacifics were great locos, they certainly did their job in pulling heavy trains. But his novel chain-driven, oil-bathed, valve gear was less successful. It had required a lot of labour to sort out the teething problems, and it continued to cause high coal, oil and maintenance costs to such an extent that BR found it worthwhile to spend money to rebuild the locos with traditional motion. By all accounts Bulleid would have preferred their being scrapped than being rebuilt. It thus seems he couldn't accept that his inventions were less than brilliant, and I suggest that he'd probably imagined Leader's new features would similarly be brilliant to the extent that he couldn't envisage any problems. It was the Chief Engineer's responsibility to sell the good points of a new design to his Directors but also to flag up their limitations and risks.

Bulleid was good at the former, he certainly sold to his management all the technical advantages planned for Leader but did he ever state that they posed a notably higher technical risk than a traditional design? The availability of such data would have helped the directors to make a better informed choice regarding approving Leader and if more than one loco should be launched. Had Bulleid "pulled the wool over the directors' eyes"?

The BR Engineering Response to Leader

Now we come to the BR decision to scrap the Leader(s). By November 1950 the senior BR engineers were aware of Leader's trial running record. Problem upon problem: 35 of the 91 of test runs deemed failures (ref 3, p. 153-155), less capable than a U class 2-6-0, excessive coal and water consumption and more lubricating oil used per mile than a diesel loco would use fuel oil (ref 6, p. 63). An intolerable working environment for the fireman. Up to 13.65 tons overweight with inequalities on left and right sides (Ref 7 p.74). Ron Jarvis correctly, and perhaps even rather generously, advised that the loco could be made to work, but would need a major redesign. By this time BR had decided on the design strategy for its Standard range of steam locos, and

One of Don Broughton's pair of amazing images showing No 36001 complete - numberplate not yet fitted - but painted black inside Brighton works. Obviously BR were not short of paint as it would be grey when it emerged the following day. We might wonder what was going through the mind of the worker stood gazing at the engine.

15

Loco		LSWR M7	Ivatt 2MT	Leader	Q1
Wheel arrangement		0-4-4T	2-6-2T	0-6-6-0T	0-6-0
Weight total	Tons (decimal)	58.8	63.25	130.5	51.25
Adhesive weight	Tons (decimal)	34.3	39.25	130.5	51.25
Weight, max axle	Tons (decimal)	17.8	13.25	21.75	18.25
Garte area	ft²	20.4	17.5	43 → 25.5	27
Tractive effort	lbf	19,755	17,410	26,300	30,080
Adhesive wt. / TE		3.9	5.0	11.1	1.8

Leader's design grate area was reduced due to the subsequent additional firebrick lining.

Leader was very different. Even if any worthwhile benefits were eventually to be demonstrated by a redesigned Leader, the concepts would be too late to incorporate in the BR Standard Classes but could be available for their steam successors. Riddles and Cox, weighed the cost and effort needed to achieve a developed and redesigned Leader prototype, which even then may show little, if any, worthwhile benefit, and decided against any further spending on Leader. How perceptive they were proven to be; as it turned out there weren't any steam successors to the Standards. The decision to scrap Leader turned out to be correct and far sighted. The railways needed their meagre resources to be spent on assets that would definitely and quickly improve the running of trains, rather than on highly speculative projects. The day of steam prototype experimentation was over.

As it turned out the BR engineers did provide an excellent replacement for the M7s, in the form of the Ivatt 2MT 2-6-2 tank, a simple 2 outside cylinder design. For the technically minded the above table helps to confirm their suitability.

To Answer Kevin Robertson's Question.

Now, to reply to Kevin Robertson's two questions (ref 7, p. 77): Do you agree with the Southern Railway / British Railways allowing the [Leader] build to proceed? Was BR correct in subsequently cancelling the project?

When the SR requirement for the new engines was issued in 1944, after 5 years of conflict, the war was still being fought, with, at last, some hope of an Allied victory and peacetime returning. The company's assets had been badly maintained during the conflict and replacements were needed for worn out stock. The Southern Railway's Board needed reliable replacements of tired equipment and unproven adventurous new concepts definitely a 'no no'. The Leader as designed, was not what the SR needed thus should not have been built.

British Railways, in inheriting Leader, was pretty generous in its continuing support of testing the loco which proved deficient in so very many ways. It'd take a lot of money and effort to produce a redesigned Leader and it'd only be worthwhile if it were eventually to prove better than modern locos existing at the time, e.g. a good all-rounder like the LMS Black 5. Yes, Leader would be total adhesion, but were its running and maintenance costs ever likely to better traditional proven machines? Very unlikely. So stick with a Black 5's derivative, and scrap Leader. That was the right decision towards rebuilding the country's loco stock.

May I conclude with words of two very able railway engineers?

Colin Boocock: "The 'Leader' class project is a classic example of what can happen when a management team loses sight of the original objective of a project and finds itself led into a situation that has taken on a life of its own, growing away from what was really wanted as it proceeds." (ref 4, p. 113)

Stewart Cox, on postulating if nationalisation had not occurred:

"All that is reasonably certain is that as long as he was in harness, Bulleid's volatile mind would have continued to produce engineering shocks of one kind or another, and that his successor, whoever that might have been, could be assured an interesting, if less dramatic, time of getting things back on an even keel once more." (ref 8, p. 211)

References:
1. Locomotive Panorama Volume 2, E.S. Cox - Ian Alan, 1956.
2. LSWR Locomotives, The Drummond Classes, D.L. Bradley – Wild Swan 1986.
3. The Leader Project, Fiasco or Triumph?, Kevin Robertson – Ian Allan, 2009.
4. Oliver Bulleid's Locomotives, Colin Boocock – Pen & Sword 2020.
5. BULLEID Last Giant of Steam, Sean Day-Lewis – George Allen & Unwin, 1982.
6. Bulleid, Man, Myth, and Machines, Kevin Robertson – Ian Allan, 2010.
7. Southern Times No. 9, Summer/Autumn 2024, Kevin Robertson – Transport Treasury.
8. British Railways Standard Steam Locomotives, E.S. Cox - Ian Allan 1973

Even today, 75 years after the Leader project was abandoned new images are still coming to light. The two on this page are from the collection of Peter Pfidczuk and show Nos 36001 and her almost complete sister, No 36002.

In the top view No 36001 is at Lewes on 17 August 1949. The appearance of the engine at this location and on this date is a bit of an anomaly as official records refer to a cancelled test for this date. Whatever, here it is 'around' the 17 August. No 2 end (bunker end) is leading. The identity of the two men on the platform is not given but it must be said the one of the right is suspiciously like Mr Bulleid. Note also the cover is off the oscillating gear at the front of the cylinders. The BR insignia and number has also been removed from the middle panel and replaced with a number only at each end.

On the right and we have a view of the almost complete No 36002 in the car sheds at Brighton. Compare with the top view, the vacuum pipe from the engine to the bogie is missing and it also appears as if no glass had been fitted to the cab windows. Other than that she is reported to have been finished. Would she have been a better engine? Unlikely. Perhaps there had been less haste in her construction which might have made a slight difference but that was all and she would certainly have displayed most of the same weaknesses as her older sister.

Schoolboy memories of the 'Belle...

Working away on something non-railway related, a 'ping' from the computer and the above arrived courtesy of Roger Geach. It is a location I personally know well although I will admit the angle from which the view was taken is totally new - proves a so a new perspective may be found even nearly six decades after the view was taken on 29 October 1966. In case anyone had not gathered, this is the 'Down' 'Bournemouth Belle', the actual location the top of the cutting very close to the north end of Wallers Ash tunnel between Winchester and Micheldever - hence also the rising sides of the cutting. There are certainly no roads in the vicinity and in consequence the photographer would have had to make his way on foot some distance from the nearest lane The down 'Belle' was an early afternoon service and so the time would currently be around 1.30pm - some few miles south and I could see it from the window at school and was frequently told off for turning my head to look, such chastisement perhaps accompanied by the words, 'Robertson will you pay attention - trains will do you no good at all'. Decades later I wish I could reply, 'Sir .... . ' The locomotive is also a personal favourite, I had two, Nos 35004 and 35012. This is the latter, certainly not in 'the prime of life' any more, indeed with just six months more to live and leaking a bit of steam from the front end. Underneath all that grime though will be gleaming green paint, perhaps the last time this was seen to all being in the spring 1964 when it was used to haul a special train of dignitaries from Waterloo to Southampton Docks at the time A4, No 60008 *Dwight D Eisenhower* was presented to the National Railroad Museum of America. During the acceptance speech, the American delegate acknowledged both Nos 60008 and 35012 adding that the latter would also find a welcome home there upon withdrawal. Such a request would fall on deaf-ears and No 35012 was withdrawn from Nine Elms on 23 April 1967 just six days after having been transferred to the London shed from Weymouth. It would be reduced to scrap at Cashmore's, Newport six months later. Even at this late stage nameplates - well at least one - are still attached whilst the route discs are pristine white. The engine also has AWS fitted, the battery box visible beneath the front step. The track work is freshly ballasted, much of the Bournemouth line similarly dealt with ready for the introduction of the forthcoming electric service in 1967. Loose rails are also visible between the 'four-foot' of the Down line and in the 'six-foot'. In the case of the latter these are short sections, those in the 'four-foot' cannot be ascertained but it was practice at the time to replace shorter lengths with CWL (continuously welded rail). White ceramic pots to support the third-rail are also in position these are sited on every fifth sleeper and have yet to alter to dirty brown. The top of the third rail is rusted so no power yet; that would come in January 1967 when this section was energised. In the background is Wallers Ash signal box which controlled the main line and loops at this the south (east) end of the four track section from Weston. Upper quadrant signals dominate the scene near the signal box and these too are on borrowed time as the line from Basingstoke through to Eastleigh was converted to Multiple Aspect Signalling just a few weeks later. Few photographs appear to exist of Wallers Ash again likely due to its inaccessibility - who can blame the photographers of the day for standing on platforms or more easily accessed paths (and who can blame them). Today there is no more 'Bournemouth Belle'. No more do BR Mk1 BG vans 'top and tail' the formation - for a time Western Region BGs in a more suitable 'chocolate and cream' were used and matched the formation much better. The scene too has altered, almost beyond recognition. What remains are the rails but that is about it, the cutting sides now full of trees that have grown up since the end of steam and are rarely cut back; slipping here in autumn due to leaves is commonplace. The loops still exist but their position readjusted due to weakness in an under bridge. I am also sixty years older. *'Sic Transit Gloria'.*

The E1R tank engine class

Rebuilding or modifying existing steam locomotives to effect improvement or to suit a new need was something undertaken by most designers during the course of their tenure.

On the LSWR for example Mr Drummond's big 4-6-0s were hardly a success and yet when rebuilt by Robert Urie there was a transformation. Thirty-plus years later Ron Jarvis devised a re-design for the original Bulleic pacifics, these two examples best described as being somewhat more radical than the three examples pursued by Richard Maunsell during his own time in charge.

These three types Maunsell was involved with were the conversion of the 2-6-4T 'River' class tank engines into 'U' class 2-6-0 tender engines, the conversion of the 'L' class 4-6-4T task engines into a new class of 4-6-0 tender engines designated the 'N15X' type and finally the conversion of ten members of the existing 'E1' class of 0-6-0T of tank engines into 0-6-2T type machines.

It was during the mid 1920s that two coinciding events took place more than one hundred miles apart. The first was a realisation that the Southern Railway possessed in excess of its requirement of 'E1' class 0-6-0T engines and the second that some additional tank engines were required in the West Country for the new Torrington to Halwill line currently being worked by the two former Plymouth, Devonport & South Western pair of 0-6-2T engines Nos 757 and 758. These had been built for the PD&SWR by the contractor Hawthorn Leslie & Co. These engines had a maximum axle load of just 16 tons and with no comparable engine available from existing stock an approach was made to the builder for an estimate for further engines - the number not specified. Harry Holcroft states it was he who worked out a scheme to modify engines of the LBSCR E1 class to which Maunsell then approved. Savings were made by utilising pony trucks previously obtained as parts from Woolwich Arsenal. In this way a saving of £12,500 was calculated compared with new engines from outside.

No 32124 at Brighton shed on 25 March 1955 and fresh from overhaul at the nearby works and no doubt shortly to commence its long journey west. Originally built in August 1878 it was already 77 years old but would achieve a further five years in traffic being one the last three to survive into 1959. *Transport Treasury*

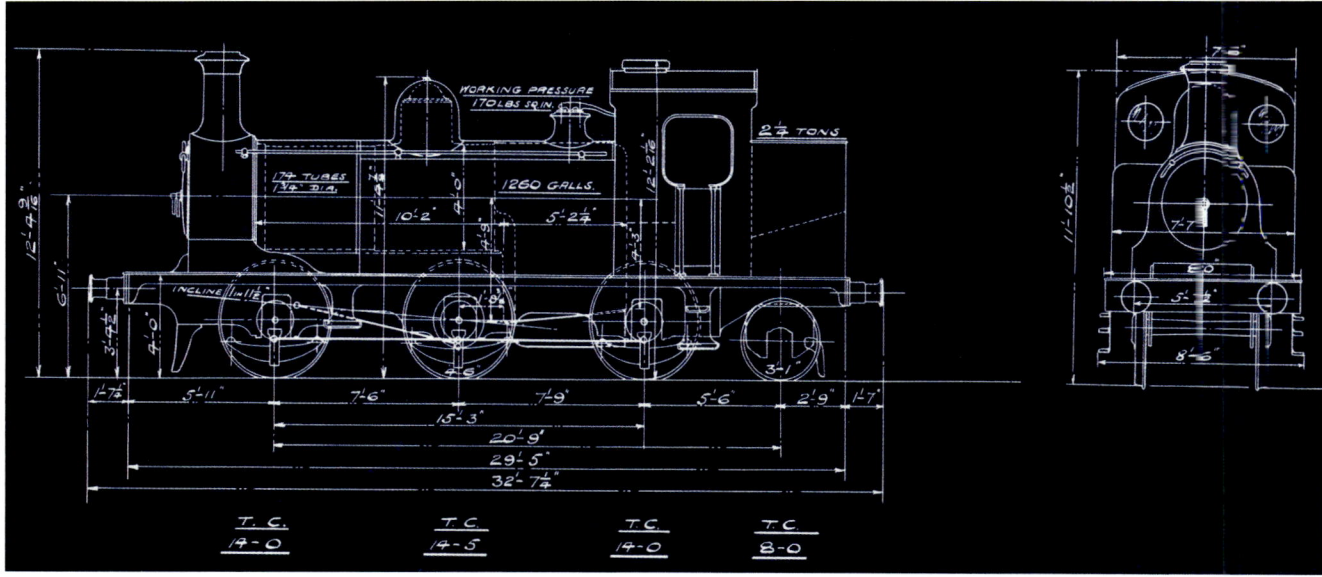

Above: Having arrived at Exeter Central from a banking turn. No 30124 awaits its next move. Despite the presence of the large weights on the driving wheels. this was one of the non-balanced engines. *David Williams / Transport Treasury*

Opposite: For comparison purposes, this is an original 'E1' 0-6-0T. Apart from the wheel arrangement, the most obvious difference compared with the rebuild type is the cab. No 2141 had been built in March 1879 and lasted into BR days being withdrawn in October 1949. Four members of the 'E1' class were also used on the Isle of Wight. *Transport Treasury*

The ten 'E1' class engines selected had the numbers 94, 95, 96, 124, 135, 608, 610, 695, 696 and 697; the last five referred to already renumbered by the Southern from their original LBSCR two-digit identification.

The new design had the need to increase water capacity above the existing 900 gallons together with a similar increase in the 1¾ tons coal carried. This was achieved by extending the frames rearwards under which was now placed a 'N' class pony truck. A larger coal bunker, well tank together with revised side tanks were added and in between a new cab visually similar to a South Eastern style. The revised capacities were now 1,260 gallons of water and 2¼ tons of coal.

The ten conversions took place at Brighton between May 1927 and January 1929. Although intended for the West Country, three freshly converted engines Nos 610, 696 and 697 spent a short time at Fratton engaged on shunting and local workings including empty stock movements to Eastleigh. Eventually all ten moved west and where one omission during rebuilding became apparent; the need to rebalance the driving wheels in consequence of the increased overall weight and which had quickly resulted in

complaints from would it be believed passengers and not the actual locomotive crews.

The result was that half of the type, those engines deemed to be regularly employed on passenger work would be attended to and which were Nos. 2094, 2095, 2096, 2608 and 2610. This work was undertaken between May 1936 and August 1937. The five altered engines were based at Barnstaple Junction for Torrington and Halwill services, the unmodified engines used for shunting at Fremington Quay or pilot and local goods workings from Exmouth Junction. They also replaced G6 0-6-0T engines on banking duties at Exeter. Two of the rebuilds survived being stationed at Plymouth in WW2,

All ten rebuilt engines entered BR services and were renumbered accordingly. Liveries varied between lined back and plain black, the latter perhaps expected to be applied to those not working passenger services but that was not the case as 'balanced' engines Nos 32096 and 32610 were always in plain black. Most repairs were undertaken at Eastleigh but in 1954/5, pressure of work meant three were sent for repairs to Brighton, one of these three 'borrowed' whilst in Sussex to work a permanent way train to Tunbridge Wells.

#March 1955 was the start of a crossroads for the rebuilds. In that month No 32124 had been given a full overhaul but just two months later in May, sister No 32694 was withdrawn.

The influx of new Ivatt Class 2 2-6-2T types from the Spring of 1953 meant their days were also numbered with the new engines taking over trains on the Torrington - Halwill line. This change saw the class now split between Exmouth Junction for banking duties and Plymouth, three engines allocated to the latter location. Four of the class did not see out 1956 and two more went in 1957. This left three working at the start of 1959 with No 32697 the final engine and which left Exmouth Junction under its own steam destined for scrap at Ashford in the Autumn.

It was not perhaps totally the end, for at Salisbury No 32697 was borrowed to double-head a permanent way train to Eastleigh after which it recommenced its journey east only to have this interrupted again at Eastbourne where it found work on carriage and yard shunting for a month, this from mid October to 22 November. It finally reached Ashford on 4 December and was broken up in January 1960. Even then the boiler was salvaged and after repairs was sold to a market garden in Sittingbourne as a somewhat large greenhouse heater.

The late Harry Holcroft in his 'Locomotive Adventure' books published by Ian Allan (1965) adds some further detail to the conversions as well as seeking to promote his own involvement in the original idea. Bradley is perhaps wise to not enter that particular discussion.

Holcroft comments that one of the duties for which the conversions was required to undertake was to take a train which included the through coaches of the 'Atlantic Coast express' from Torrington to Barnstaple Junction. Here they would be added to the through caches from Ilfracombe and that consist then taken through to Exeter. The Torrington train was on a tight schedule meaning at times it was running at 50mph and the 4' 6" wheel were rotating at five times per second. It was this that had resulted in the passenger complaints and was best described as at extreme example of the backwards and forwards oscillation so often experienced in the first few coaches of a train hauled by a 2-cylinder engine. A simple solution, but at the expense of local passengers, would have been to place the through coaches at the rear of the train but this was rejected as likely to cause shunting delays at Barnstaple Junction.

He then goes on to describe in detail the cause of the rough riding in the train. This was due to the original design where Mr Stroudley had the outside crank pins coinciding with the inner cranks instead of the more usual arrangement of being opposed by 180°. Stroudley's justification was there was then less wear to the axle boxes but it did require very large balance weights be provided on the engines - as may be seen from the photographs.

An experiment was tried on one engine - number not given - with the crank pins repositioned to be opposite the inside cranks. The results were favourable but Holcroft comments that only three engines were converted - he is also of the impression that there were 11 and not 10 conversions.

Left: SR No 2135 at Exmouth Junction shed on 5 August 1938. This engine had been converted ten years earlier. It presence here indicates it was one of the type regularly used for banking duties. *Transport Treasury*

Opposite top: Rebalanced No 32608 being admired at Hatherliegh - probably more admirers than actual passengers! The lack of a headlamp will be noted. (Further views of the stations between Torrington and Halwill Junction appeared in ST9.) *Neville Stead / Transport Treasury*

Opposite bottom: No 32135 on banking duty at Exeter St Davids, 7 August 1954. The tanks on either side of the rear framing are vacuum reservoirs. *Transport Treasury*

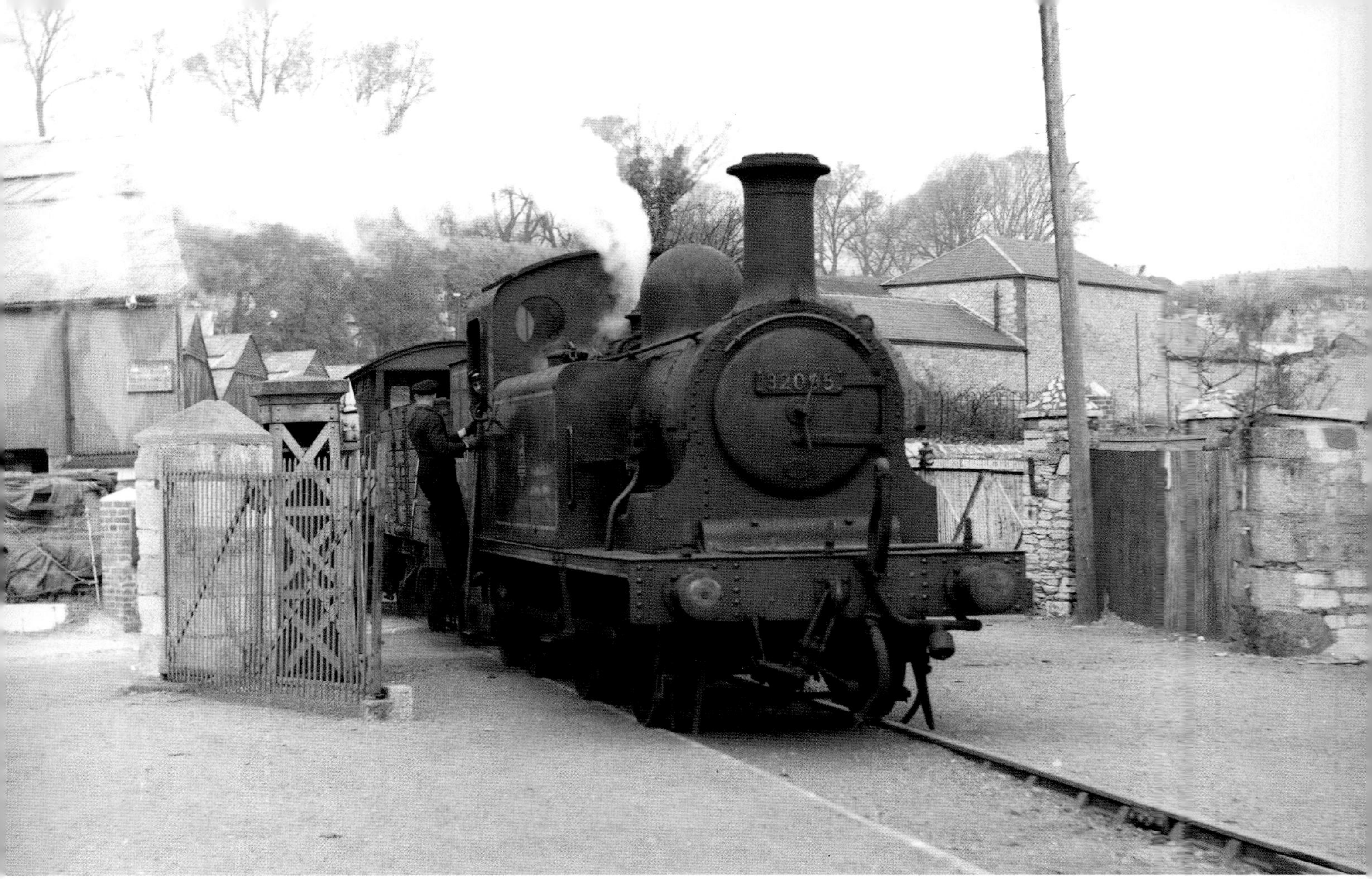

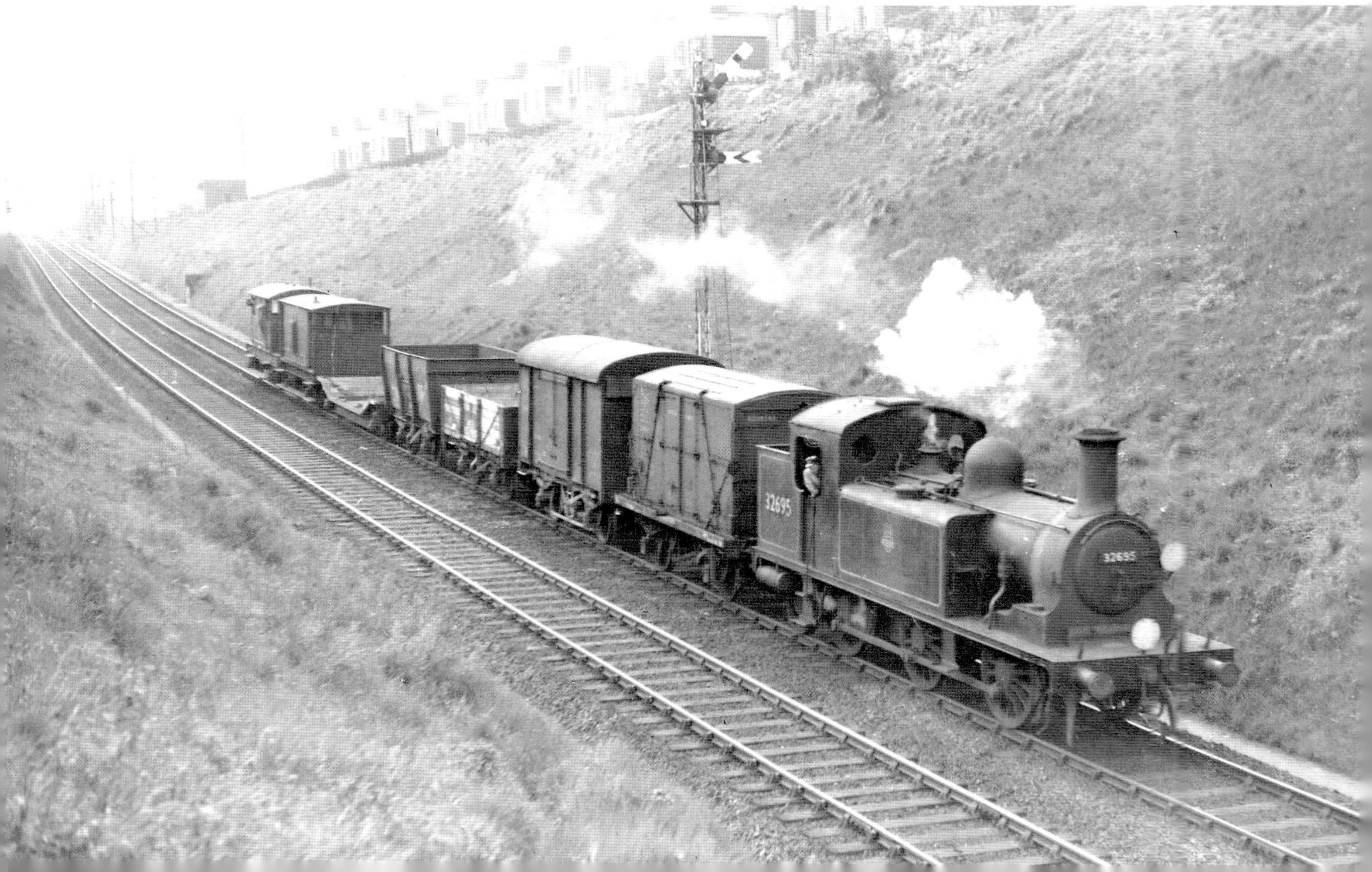
32695
32695

Opposite top: Rebalanced No 32610 shunting the Ocean Quay at Plymouth, 22 April 1954. *A Lathey / Transport Treasury*

Opposite bottom: No 32695 with a short freight near to Exeter Central. The disc code indicates trains between Exeter Central and Exmouth Junction. *Flint & Harbart / Transport Treasury*

Above: No 32095 and a single brake van on the seldom photographed branch from Devonport to Stonehouse Pool. Access to this line was from Devonport goods yard and a long under bridge passing under the goods shed. The main line could then be accessed at Goods Junction. The single disc also applied to trains on the Cattewater branch from Friary. Services officially ceased in 1970 although the last train had run in 1966. *A Lathey / Transport Treasury*

LBSCR No.	LBSCR Name	Built	Renumbered		Rebuilt	Order No.	Later SR No.	BR No.	Withdrawn
94	*Shorwell*	11/1883			31/5/1927	23C	2094* (10/1936)	32094	4/1955
95	*Luccombe*	11/1883			27/5/1927	23C	2095* (12/1937)	32095	11/1956
96	*Salxberg*	12/1883			18/10/1928	32B	2096* (5/1936)	32096	11/1956
99	*Bordeaux*	12/1874	610	6/1922	14/1/1929	32B	2610* (8/1937)	32610	3/1956
103	*Normandy*	9/1876	695	6/1913	27/10/1928	32B	2695	32695	2/1957
104	*Brittany*	10/1876	696	10/1913	20/12/1928	32B	2696	32696	1/1956
105	*Morlaix*	9/1876	697	4/1915	17/1/1929	32B	2697	32697	11/1959
108	*Jersey*	11/1876	608	1/1916	13/11/1928	32B	2608* (2/1937)	32608	5/1957
124	*Bayonne*	8/1878			3/12/1928	32B	2124	32124	1/1959
135	Foligno	1/1879			26/11/1928	32B	2135	32135	3/1959

* indicates a rebalanced engine and date.

Bradley gives three sample final mileages; Mo 32095 as 1,497,037. No 32124 as 1,482,336, and No 32135 1,120,221. These figures were comparable with original menbers of the E1 class.

Stephen Townroe's
Colour Archive
'On Somerset & Dorset lines'
Part 1 Templecombe and Bath

Templecombe was one of two important junctions mid-way on the Somerset & Dorset line (the other was at Evercreech). Geography wise, the S&D crossed under the former LSWR main line at 90°; the S&D running from Wincanton in the north to Henstridge in the south. On the LSWR line it was Gillingham in the east and Milborne Port to the west. In the earliest days of both railway the S&DJR the location was a terminus with trains from Bath terminating at a public station located approximately where the later engine shed stood. This had opened in 1861 at the same time as a north to east connection allowing interchange between the S&D and LSWR lines.

This arrangement remained until 1870 when instead of north to east connection it was replaced by a north to west connection, the former north to east connection abandoned and thereafter used as a curved siding the alteration meaning trains from the north could now use the LSWR station. When the S&D south was completed in 1863 and passing under the LSWR route, it is believed the original terminus station was closed. The final addition was a platform facing the single line S&D line just north of and just prior to the S&D crossing under the LSWR.

At no time then was there ever a triangular junction at Templecombe although there were certainly plans for this drawn up on at least one occasion and which involved reinstating the original north to east connection.

Most S&D train services were thus compelled to either run in and reverse out, or reverse in and run out if the service required access to the main Templecombe station. From photographs the layout here might appear to be excessively complicated but in reality it was straightforward even if three different levels were concerned.

Opposite: This is looking north with SCT positioned close to the two lines leading from the S&D to the LSWR station. In the centre is the single 'main line' of the S&D running north to south and on the right the BR built engine shed and various sidings. 2-8-0 No 53802 is just in view. June 1952.

Above: A Johnson 3F reversing back towards the shed again in 1952.

Opposite; Here 4-4-0 No 40697 is approaching from the north and about to pass the bracket signal which will direct the train either on to the single line south or to head toward the LSWR. (Although obscured by steam this service, a local working, will reverse to take the line to the LSWR station.) Close observation reveals headlamps over the right hand buffer and under the chimney; this was the one of the special lamp codes that applied to the S&D and indicated to signalman it was a train calling at Templecombe. The signals and connections here were controlled by Templecombe No 2 (Junction) signal box and from where SCT has taken his picture - notice the rodding run in the foreground. The signals numbers left to right are 23, 24, 25, 26 and on the far side 8 (push) and 18. On the right a p/way trolley hut is built into the embankment. After having been dealt with at the station, the Templecombe pilot will attach to the rear and haul the train backwards to Templecombe Junction after which it will detach allowing the service to continue south. The whole was a convoluted affair, expensive in time and manpower the cost of which no doubt contributing to the overheads attached to the working expenses.

Above; This time we see the signal box referred to. Taken from the sidings of the lower yard the difference in levels is appreciated. First there is a 4-4-0 standing on the connection from the LSWR station, then in the centre another 4-4-0 approaching signal No 6, an upper quadrant which is the 'off' position. The train is a Maunsell '3-set' with at least one van tacked on behind. The train will run forward and come to a stand ready for the engine on the high level line to drop down and attach to the rear ready to haul it to the former LSWR station. Notice too the individuals walking in the 'four-foot' and accompanied by two railwaymen. We are not told the date but perhaps with SCT present as well this was an 'official visit' - to view this expensive to operate station deep in the heart of rural Somerset. On the right piles of wooden sleepers have accumulated alongside the siding whilst in the foreground are various allotments mostly complete with produce and no doubt cultivated by the local railwaymen. The view is of course looking north with the remaining signals, as seen in the view opposite, of the lower-quadrant type. Looking at this image and the other views of Templecombe, the number of sets of rails, the infrastructure generally and the whole might almost appear permanent and unchanging; as indeed it was for almost 100 years, but then on that fateful day in 1966 all would change and what had been consistent was simply cast aside. SCT does not give dates for either of these views.

Part 2 of 'On Somerset & Dorset lines'
featuring the intermediate stations
will appear in Issue 15.

Above: In this view we see two former LMR locos which SCT has identified as on the left as 3F No 43194 - an original S&DJR loco, and on the right, a 4F No 44102. The 3F is coming off the single line from the direction of Blandford; might it be the same engine seen reversing back on to the shed on page 27? In which case it still has the letters 'L M S' on the tender and this is May 1952. The 4F is waiting at the signal on the line leading down from the Southern station.

Opposite top: Transfer goods between the Southern and the S&D. The engine seen from the rear could well be the 4F referred to in the previous view. With an engine at each end it would seem this is most likely a move back up to the Southern even if the crew are looking forward; the single white lamp on the rear would similarly appear to confirm this. Might that even be another 3F at the front? The tender of the 4F appears well stocked with coal and the crew also have a storm sheet erected. On the rear of the tender it is possible to read the letters 'L M S' on the works plate. Notice also the lid to the water filler, a simple 'dustbin lid' type item, austere yet functional. Counting the wagons of the train and we would appear to have in the order of 25 vehicles of mixed type and including two brake vans. Towards the front there is also a wagon carrying what appears to be a 3-ton Morris commercial. The view was taken in May 1952.

Opposite bottom: Again taken at Templecombe No 2 but without further covering detail. Ignoring the presence of the workmen (and their shovel nearest the camera), what appears to be a $F is pulling another train on the reversible passenger line up towards the Southern station. The train engine is of course at the opposite end and after necessary station duty it will depart again forward towards Evercreech and Bath. From the storm sheet, might the engine seen from the rear even be the same one in the previous view. Pause for a moment also to consider a passenger travelling on the S&D for the first time - in either direction. It seems doubtful the booking office clerk, ticket collector or guard would inform passengers of the manoeuvres at Templecombe and in consequence from venturing forward to then stopping to start reversing must have been alarming to some - especially those who prefer to travel 'with their back to the engine' or whatever. Turning now to the signals facing the camera on the left hand side, these are the starting and shunt signals for trains on the left hand line to head towards Templecombe No 3 box.

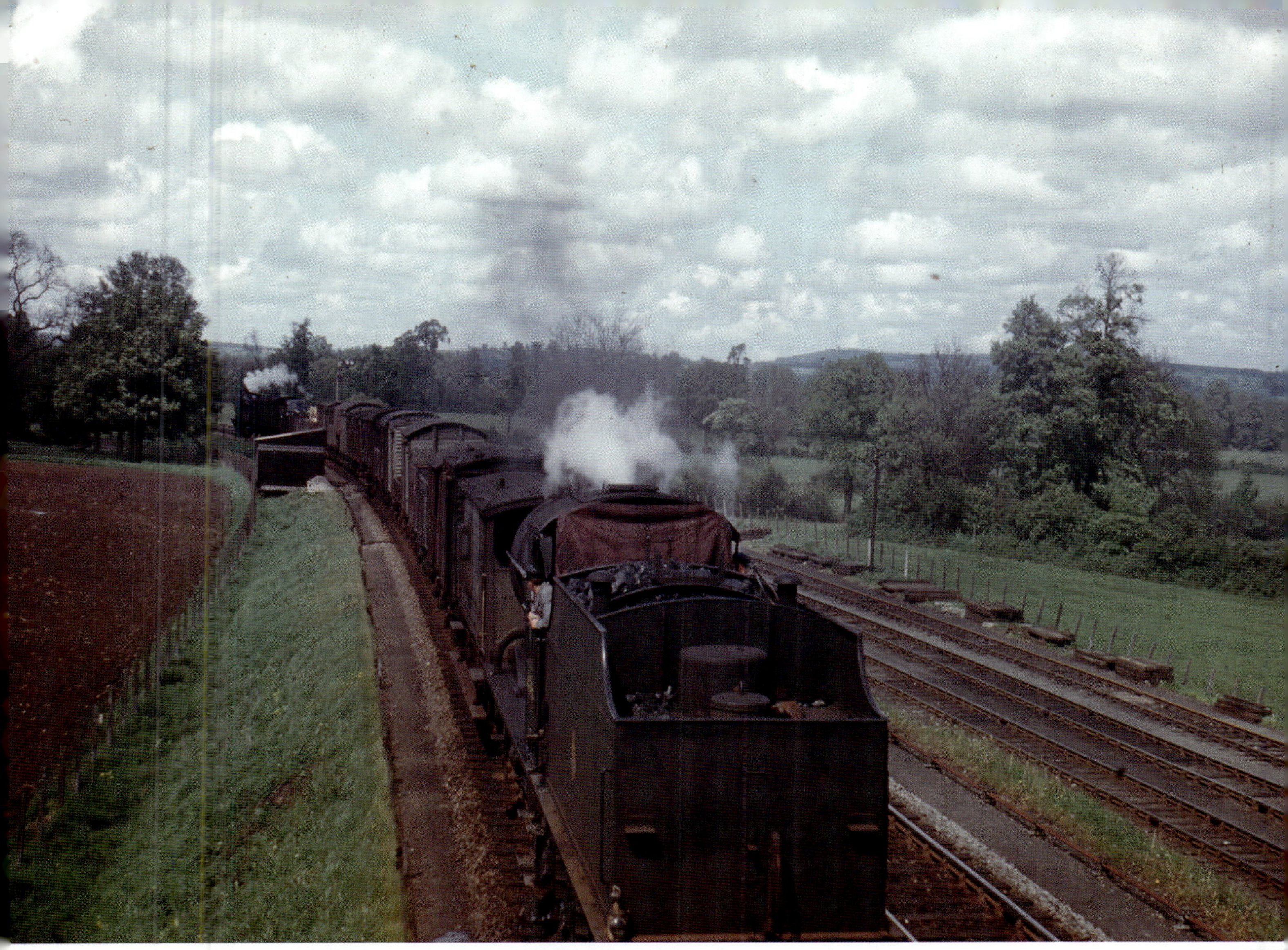

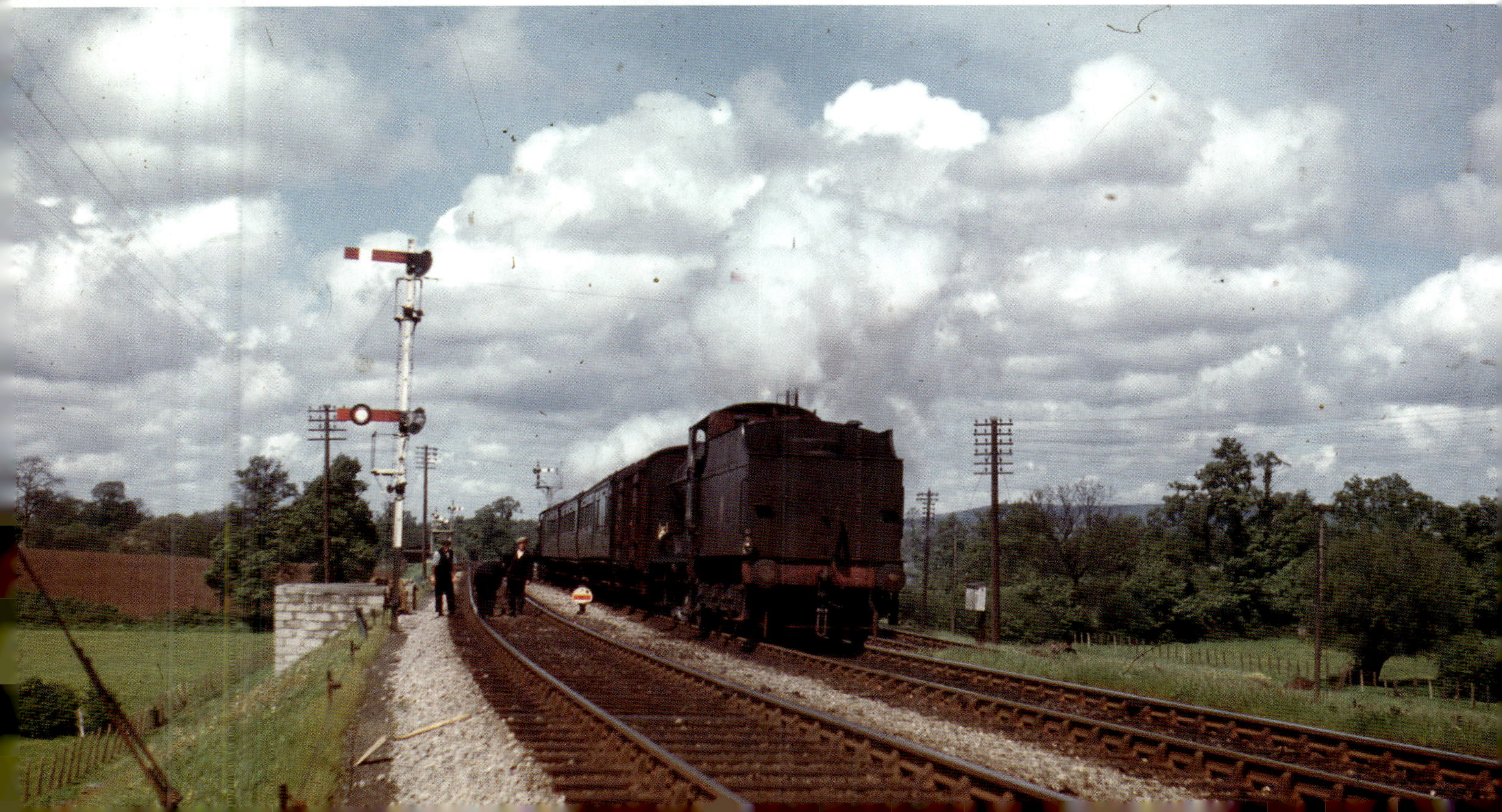

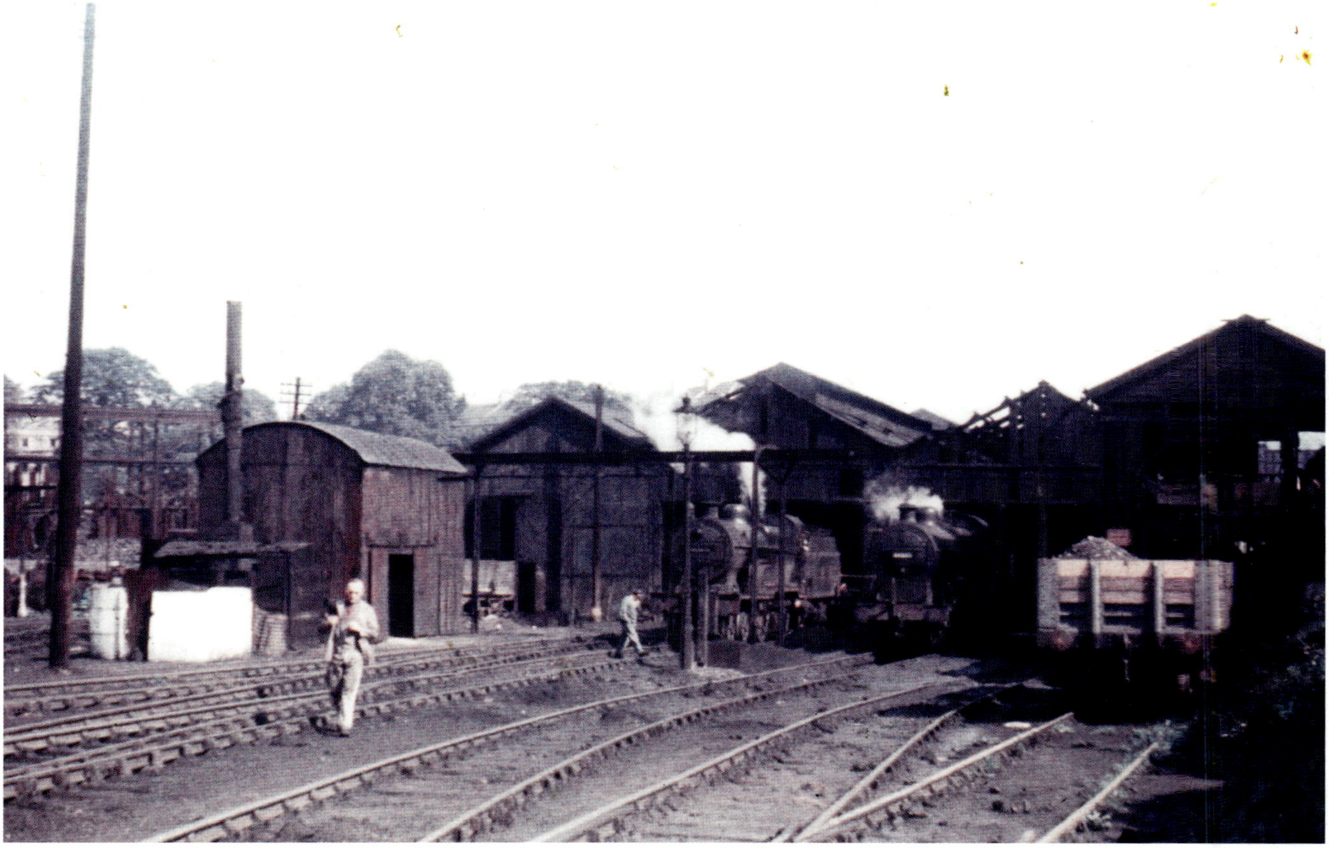

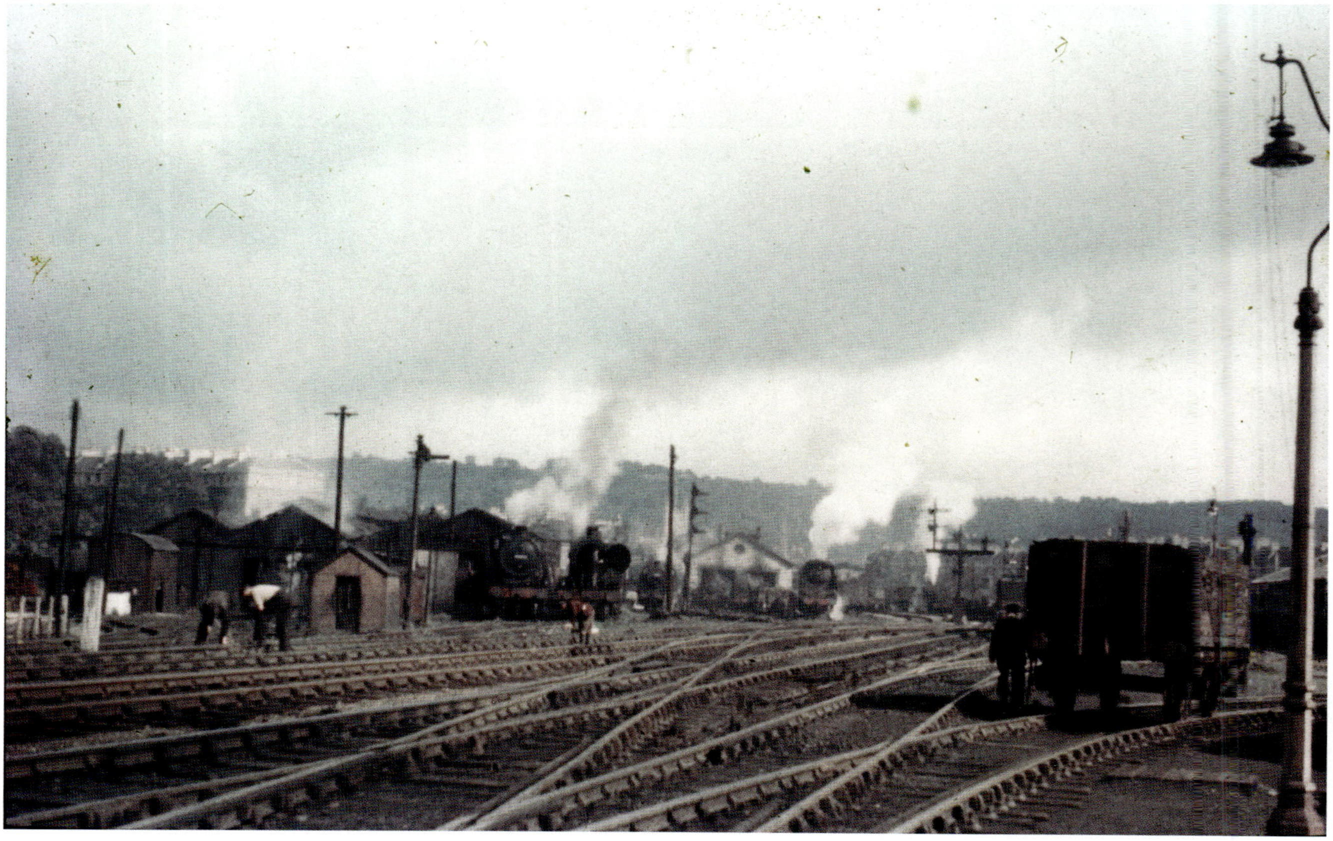

Opposite top: A view of the S& D shed at Bath Green Park in July 1952 with two locos visible; a 4F and a 4-4-0 the latter having 'L M S' on the tender. In the left background are some sawmills, the corrugated building to the left - he one with the chimney - the sand furnace. A turntable was off to the right which is where the coaling stage was also located. Various offices and mess rooms were scattered around the site having grown piecemeal as the site developed in the late Victorian and early Edwardian eras. (Peter Smith in his 1978 OPC book 'Footplate over the Mendips' book, provides some excellent recollections from footplatemen who were clearers and fireman at Bath over a century ago. Might the man walking towards SCT be either a locoman or even a photographer?

Opposite bottom: There were to loco sheds at Bath, the S&D shed and the smaller but more substantial stone built Midland shed. The latter is seen straight ahead and with a Bulleid pacific blowing off steam - the line to the station continues to the right with four sets of rails leading across the River Avon to Green Park terminus. For many years the two sheds were run as totally separate entities, the Midland shed catering for locomotives working to Bristol via Mangotsfield. Like Templecombe this was a railway conglomeration that appeared permanent especially when considering the number of lines. Men started their working life here and would retire from here ans yet nowadays with nothing to show for their endeavours. At bath there was even a Railway clearing House office to deal with the transfer of traffic between the companies. On the extreme right the wagon is on a siding that ran close to some rail served cattle docks and would terminate close to Lower Bristol road on which ran the Bath tramway until 1939. Out of sight to the right were a number of stables for shunting and carriers horses.

Above: LMS Class 2 No 41242 on a shunt move at Green Park. There were two platform faces here, that to the left normally used as the arrival line and the one on the right for departures, although both were signalled in either direction, a useful feature at busy times. One the left is one of two grounded coach bodies, presumably used as storage although this cannot be confirmed. The station was gas lit. This particular view was taken in August 1951 and when the engine was two and half years old. Built at Crewe it was immediately allocated to Bath Green Park but moved to Templecombe in May 1952 and which is where it remained until withdrawn on 30 April 1965. It would not survive long and was scrapped four months later. The station at Bath Green Park was fortunate to survive closure and after temporary use as a car park has been re-purposed part as s supermarket site and part market / cafe culture.

April 1950 footplate view from the cab of 'Black 5' No 44835 being piloted by Southern 2-6-0 No 31625. This was the month when motive power supervision became the responsibility of SCT from Eastleigh although he was wise enough to let things remain as they were even if instructed by 'higher authority' to try various Southern types on the line - without really improving on the existing stock. The train is coming down the grade and about to join the Midland main line at Bath Junction. It will already have crossed over the GWR main line and is seen here passing over the Lower Bristol Road. It was at this junction in 1929 that a train careered out of control down the grade the crew having been overcome with fumes in nearby Combe Down Tunnel. Towering above the engines is one of three gasometers of Bath Gas Works, not surprisingly there was rail access here from the other side of the line. On the right the rails lead to green park; there was never any direct access towards Bristol and reversal of direction was required to gain that route.

Next time in the **SCT** colour archive: 'Whitchurch 1954'

Part 2 of 'On Somerset & Dorset lines'
will feature in Issue 15

Peter A. Harding

of the railway from the original bill of 1864, so politics would again intervene in the 1960 when British Railways proposed closure citing an annual loss of £26,000. Slightly unusually the Transport Users Consultative Committee argued against closure on the grounds of usage. It was a strong case but the man in whose hands the decision laid was Ernest Marples.... .

Closure then took place in late October 1961 and despite valiant efforts by some to operate the railway as a private venture that idea too fell by the wayside with part of the trackbed near to what had been Brasted station now covered by the M25.

Peter Harding's book covers all these points and more. It is beautifully written and amply illustrated (60+ illustrations and track plans) and will easily consume some spare moments accompanied by one's favourite beverage or tipple

'The Westerham Branch line' is privately published by the author at 'Mossgiel', Bagshot Road, Knaphill, Woking, Surrey. GU21 2SG. Softback, 32 sides with paper covers printed on art paper. 209mm x 146mm portrait. Excellent value at just £5.00. ISBN 978 1 03691 6 24. Copies are available directly from Peter if needs be.

It is difficult nowadays to find a stretch of railway that has not been the subject of forensic analysis often more than once. The literal 'sleeper by sleeper' histories have their place but so too does a book that tells the reader what he or she needs to know (and probably did not know what that was beforehand) without leaving them yarning as page after page is filled with perhaps parliamentary and legal battles.

Books by Peter Harding fit neatly into the 'what you need to know' category and none more so than his latest offering on the Westerham line.

We will not be stealing Peter's thunder if we briefly state the branch line opened in 1881 from a junction with the South Eastern at Dunton Green. The original plan had been for a railway from the latter location through to Oxted but the final section was not built. Had it have been it would have provided for a useful connecting railway instead of being a branch line through rural Kent to a simple branch terminus.

In the same way politics had delayed the construction

From Chevening Halt the line contined climbing a 1 in 76, 148 and 74 and having passed under a road bridge, which carried a minor road (now known as Chevening Road) leading to Sundridge and then under another bridge which carried the road to Combe Bank (sometimes spelt Coombe Bank).

Having now reached the lines summit, it then dropped at 1 in 217 and 515 and passed under yet another bridge, which carried Brasted Hill Road, before arriving at Brasted Station, (later reduced to a Halt) which was 3 miles 6 chains from Dunton Green.

The rather attractive wooden station building was situated on a single platform, on the 'down' side (south) of the line while a small three lane goods yard was situated behind the station building and was entered from the Dunton Green direction. In the earlier days of the line, Brasted had its own signal box but, this was later removed and the sidings were then controlled by a 3-lever ground frame.

Although Brasted never had its own goods shed, it did have a grounded goods van body which was thought to have been from a former London, Brighton & South Coast Railway goods van, although later reports think it was built for the SER in 1918 by the Metropolitan Railway Carriage Wagon Co.

On September 19th 1955 the station was down graded and became an unstaffed 'Halt'.

Unfortunately, Brasted Station/Halt was over half a mile from the village and was reached by a long sloping approach road which lead up to the station from Station Road.

In September 1881 William Tipping had arranged for this long sloping road to be built with local money after the SER agreed to pay £100 towards the cost.

Site of Signal Box

To Chevening Halt

To Westerham

Station Building

BRASTED STATION (LATER HALT)

BRASTED HALT

Approaching Brasted Station on October 28th 1961 with the goods yard on the left.   Ian Nolan

The Transport Treasury archive of literally hundreds of images regularly throws up surprises. If the requirement is for a particular locomotive at a specific location that is not always possible but there are more than enough other compensations. Here we present a few recent 'finds' some from collections have not be listed and images that had certainly not previously been scanned.

Opposite top: No 34035 Shaftesbury with its strange front fairing seen here at Ramsgate on 10 May 1960 in charge of the 'Man of Kent' service. *Dave Clark*

Opposite: Chestfield and Swalecliffe Halt. The Bullied creating a smokescreen is unidentified; we may assume though the regulator is closed. The train is a down service in the direction of Herne Bay. The collection of bicycles is worthy of a second glance. *MC10012C*

Above: Tulse Hill exterior and outside news-stand. Here the notice warns against bicycles but clearly it does not apply to all! Indeed notice on the rear mudguard of the bicycle the white painted section; a WW2 requirement and still apparent on a number of machines twenty years later. Southern Railway signs dominate, the net curtains indicating this was the areas of living accommodation. Compare that with the window bars protecting the booking office at the far end. *MC10009T*

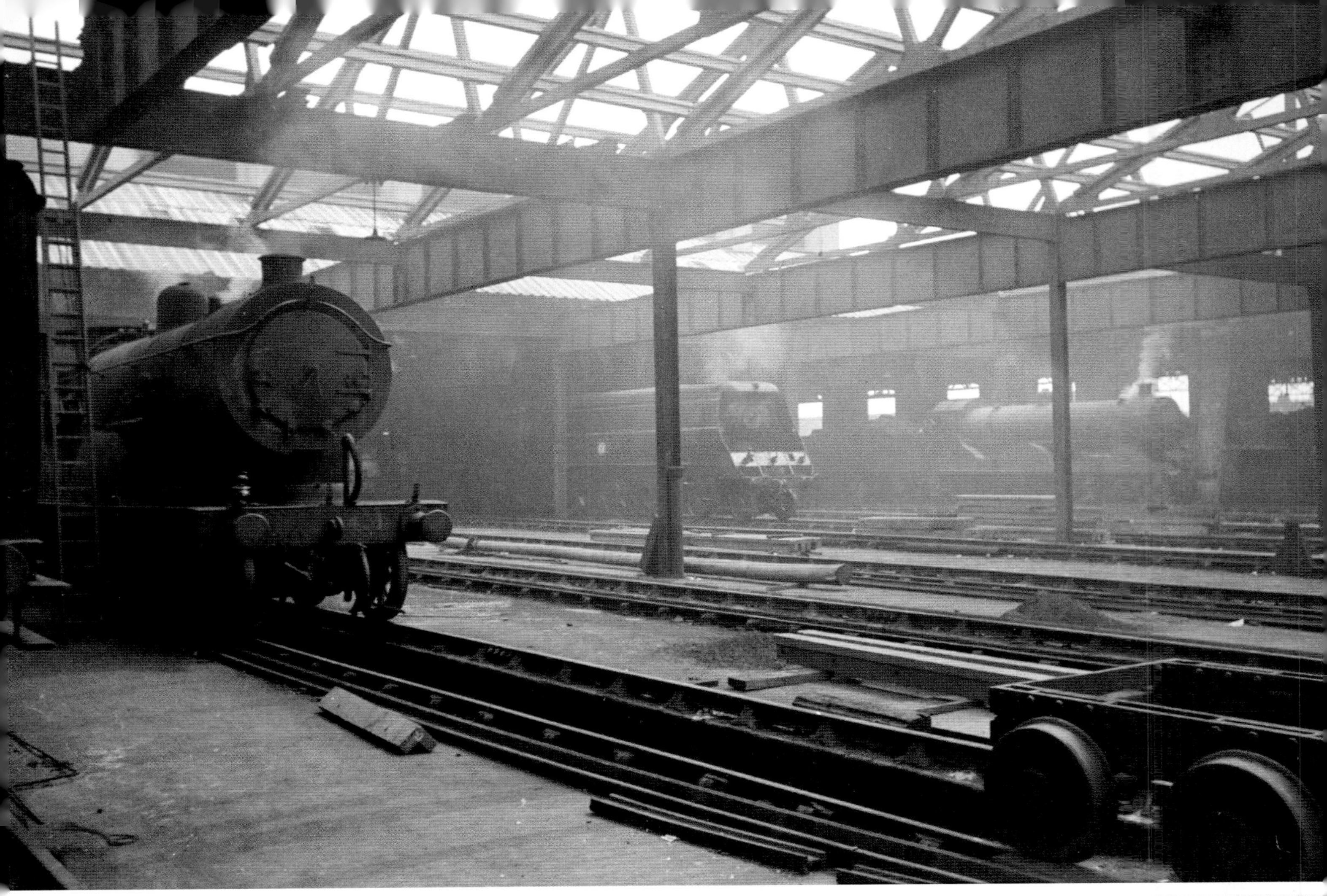

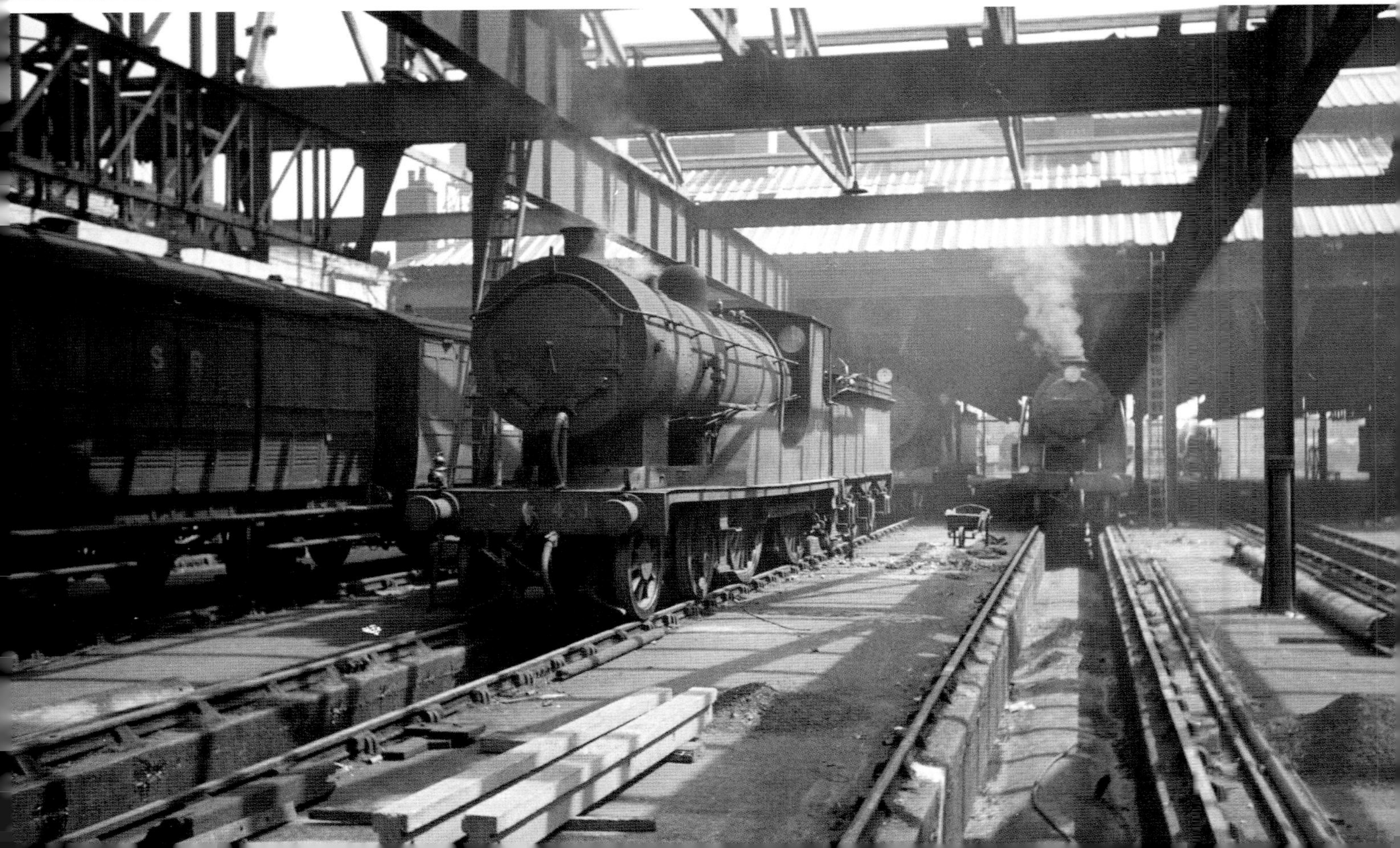

Opposite: Two atmospheric views of the interior of Nine Elms shed by Roy Vincent. Both were taken in July 1947 with SR lettering dominant on the engines seen together with what is almost certainly a mostly glassless roof - wartime damage. In the top view we have what is an L12 4-4-0. No 431, this engine was based at Nine Elms in 1949 and after transfers to Basingstoke and Guildford was taken out of service in October 1951. In the back ground is another Drummond 4-4-0 and also a Merchant Navy, most likely No 21C12 *United States Line*. From the presence of the ladder it appears structural work is still being carried out. The 4-wheel trolley was used to move heavy items around; coupling rods, buffers etc. In the lower view Roy has moved across slightly to show No 431 from a different angle with another 4-4-0 behind and a 4-6-0 of some type. The open roof shows up well as do the pits and general detritus. On the extreme left are likely vehicles from the breakdown crane.

Above: Billinton B4X 4-4-0 No 2072 seen at Victoria in August 1937. With a tender full of coal this would be a departure and very likely a special working - Epsom Downs perhaps? No 2072 had been built for the LBSCR by Sharp Stewart and entered service in September 1901. A total of engines were in the class eight of which were built at Brighton. For around a decade they handled the heaviest and fastest trains on the LBSCR until that mantle passed to the new Atlantics and large passenger tanks. For a time they were cascaded to lesser duties until a number were rebuilt with 'K' class boilers but which resulted in a need for new frames. To save costs the original motion and inside cylinders were used but with the new larger boiler no improvement could be made to the valves and exhaust arrangements and they never really outshone the unconverted engines. Not surprisingly withdrawals started in the 1930s and had it not been for WW2 no doubt all would have gone during the time of SR ownership. Instead six original B4 class engines and 12 of the converted B4X passed to BR but all had gone by 1951. No 2072 was originally LBSCR No 73 and had once carried the name *Sussex*. The last three survivors, including what was now No 32072 were withdrawn on 1 December 1951, records show No 32072 was scrapped in the same month.

Overleaf: An undated view of Victoria and almost certainly from the 1950s. The 2-6-4T is announcing its presence as well as being observed by the smallest of the children. As with so many images of the period, the engine is only part of the scene; the 'Birdcage'. season ticket sign and use of the word 'SHEWN', poster adverts and the finger board. Have we missed anything...?

16

SHEWN

OTTED and EAST GRINSTEAD

ADELPHI

Huntley
& Palmers'
Biscuits

OREAL
PASTEL

PE WORTH
HO SE

Que sertil à un homme
de gagner tout le monde
s'il perd son âme ?

PARHAM

To kill a railway: the Ilfracombe line

Followers of our sister publication 'Western Times' will be aware o f the commencement of a series of articles on Western Region line and station closures. This has been made possible by access to Western Region correspondence files held at the Wiltshire and Swindon History Centre at Chippenham and where, by prior appointment, it is possible to examine some of the hundreds of paper correspondence files appertaining to how the Traffic Department ran the Western Region in the late 1950s through to the early 1970s.

Most of these appertain to traffic and not motive power, the traffic department material of relevant to us now as it also includes correspondence appertaining to the former Southern lines east and west of Exeter as well as the Somerset & Dorset. We may well return to the last named at a later stage (time and access permitting), but for the present we will confine ourselves to one of the most lamented of closures, that of the line from Barnstaple to Ilfracombe.

We should mention that the origins of these bundles of papers is clearly a divisional office and almost certainly Plymouth or Bristol and it is important to also note there could well be other files missing that could well answer questions or throw up further anomalies. In similar vein we often have just one half of a correspondence trail yet still sufficient to draw reasoned conclusions On the occasions where supposition has been made this is clearly stated.

We believe this is the first time the contents of this resource have been accessed for the purpose of examining the circumstances leading up to the closure of a Southern line and which for its final years came under the control of the WR.

Ilfracombe had received its railway connection on 20 June 1874. It had taken several decades and several false starts to get to this point but from this date and for almost a century there would be a train service operating from Barnstaple to Ilfracombe and serving the intermediate stations of Grafton, Braunton and Mortehoe & Woolacombe. For most of that time too

Cruelty, but perhaps truthfully, this was the representative scene at Ilfracombe for several months of the year. A terminus and a line, which would bustle with activity during the spring and summer bank holidays and likewise on Summer Saturdays. So was it a railway that deserved to be sacrificed on the altar of economy or should and more especially, could a more determined effort have been made to secure its future?

As it was. The Ilfracombe portion of the 'Atlantic Coast Express' pounding up the bank past Slade reservoir towards the summit. The engine is No 34066 Spitfire. The maximum trailing load for this class of engine between Mortehoe & Woolacombe and Ilfracombe in either direction was 240 tons. *George Heiron / Transport Treasury*

there were direct services from Waterloo and also Paddington the first signs of a contraction coming in April 1950 when regional boundary changes meant all the Southern lines west of Exeter were transferred to Western Region control. Even so day to day operation still remained with the Southern hence Southern Region engines continued to work services and through workings from Waterloo. Over the next few years, it would be a literal ping-pong situation as regards 'ownership' for in February 1958 the Southern Region regained control of its former routes (excluding those in the Plymouth area and which are irrelevant in terms of the present discussion) only to again loose everything to the WR on 1 January 1963. The WR were now the sole 'owners' of all the former routes west of Wilton and it would not be long, January 1964 to be exact, before Paddington began to seriously consider the future of a

number of its new acquisitions. We say 'consider' and use the word carefully because when the Beeching report is also taken into consideration the e was a real risk that every former Southern route, main line and branch west of Salisbury could have been lost to the railway map.

Before turning to the perils of that time let us first place the Ilfracombe line in context. A 14-mile double track 'branch' from Barnstaple Junction and which included no less than 11 level crossings including the delightfully named Duckpool Crossing between Barnstaple Town and Wrafton. There was one short single line section over the curved River Taw viaduct between the two Barnstaple stations (again we may conveniently ignore the former Barnstaple Victoria Road GWR station). On the Ilfracombe line from Barnstaple Junction as

42

far Wrafton the line was relatively easily graded but after this it began to climb for the next seven miles culminating in a cruel 1-40 with curves as sharp as 18 chains radius. From Mortehoe to the terminus at Ilfracombe was a steep descent, 1-36 in places and with ever sharper 15 chain radius curves. It was not an easy line to work and not cheap either with the number of crossing keepers required at the intermediate level crossings and the number of banking engines necessary during the Summer months.

Even so and regardless of the alternate changes of ownership there were still through trains to Ilfracombe from both Waterloo and Paddington, the prestige 'Atlantic Coast Express' including an Ilfracombe portion whilst for some years post nationalisation the stations remained fully staffed.

Times though they were changing. Car ownership small as it was in 1950 gradually began to increase and regardless of the congestion it brought it was also a wake-up call to what would occur ever more in the years to come. For the present long distance and commuter travel was safe but the former would also start to suffer as coach travel began to be seen as an alternative to rail. (How well the present writer recalls journeys from Hampshire to Devon by 'Royal Blue' with 'comfort stops' at places like Bridport and Yeovil.... . A none too pleasant experience of food poisoning from a 'British rail sandwich' on the train between Salisbury and Exeter had understandably put the grown-ups off rail travel.)

If there was a time the rot started to show it was 1955 and then again in 1958. In 1955 there was the ASLEF strike, foot passengers might find an alternative with the competing bus services in the area, some would never return, but for the movement of goods this was a golden opportunity seized with both hands by the respective hauliers; once a customer had switched to road and found there was no longer transhipping involved some would never again return to rail. The second date was when the railways were forced to implement a pay award to staff but with no additional money available from government to fund it. The results were cuts to services, the very passengers the railway needed to retain were being forced off the trains.

Unfortunately receipts for the Ilfracombe line stations for the 1950s and early 1960s are not available, if they were they would surely make for an interesting comparison with figures for earlier years – which are similarly not available.

What we do know, and here we must thank Messrs Nicholas and Reeve for information in their book on the 'North Devon Line' (Irwell Press 2010) and where on page 271 statistical information on receipts for the line are given courtesy of the 1963 Beeching Report. This indicated that at Wrafton receipts were less than £5,000 annually, the other stations returning figures between £5,000 and £25,000 (the actual numbers not given). The point was made that '…at most North Devon stations more tickets were collected than issued…' this might sound positive and proved the point the railway was a popular destination but unfortunately the BR accountancy system meant that such receipts were only credited to the issuing station, i.e. a Waterloo to Ilfracombe return was credited wholly to Waterloo.

Beeching's recommendations for the line were devasting, wholesale closure. To someone standing on a crowded platform at Ilfracombe on a Summer Saturday this must have seemed unthinkable yet according to Beeching numerous locomotives and carriages were only working on six to eight days in the summer and standing idle the rest of the time. Strange is it not that few of us from that generation can recall long lines of locomotives and carriages rusticating away on 350+ days a year. Even at this early stage we have to say Beechings figures cannot be taken as accurate as according to Nicholas and Reeve 75% of annual passenger traffic was carried in three summer months.

Later in the same year the 'North Devon Railway Report' was produced privately by the noted railway writer and publisher David St John Thomas. This report, based on an extensive survey of 7 May 1963, accepted that on that date receipts for passenger and goods traffic covered no more than 20% of the operating costs of the respective services this despite the Ilfracombe line having some of the busiest stations in North Devon.

The report made some sensible recommendations based solely on realism and not nostalgia. It said the line should be singled but with full length passing loops to cater for summer traffic, automatic lifting barriers be installed at level crossings and regular daily service reduced to four or five multiple unit diesel trains. Barnstaple Town would also be reduced to unstaffed 'Halt' status and Wrafton closed. The 'North Devon Railway Report' is not mentioned in any of the files located at Chippenham.

From Nicholas and Reeve, we also know that BR attempted to use the condition of the bridges over the River Exe at Cowley Bridge as justification for closure of – basically all the lines west of Exeter. Despite a request to the Minister of Transport this permission was

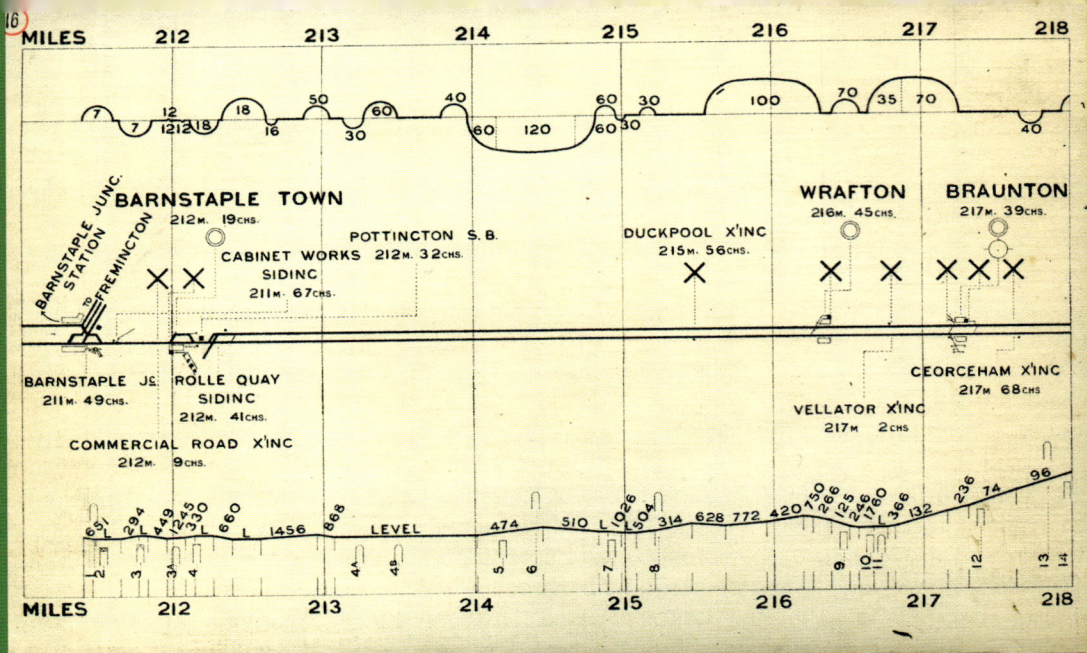

MILES 212 213 214 215 216 217 218

7 12 18 50 60 40 60 30 100 70 35 70
7 12 12 18 16 30 60 120 60 30 40

BARNSTAPLE TOWN
212 M. 19 CHS.

WRAFTON
216 M. 45 CHS.

BRAUNTON
217 M. 39 CHS.

BARNSTAPLE JUNC.
STATION
TO TREMINGTON

POTTINGTON S.B.

CABINET WORKS 212 M. 32 CHS.
SIDING
211 M. 67 CHS.

DUCKPOOL X'ING
215 M. 56 CHS.

× × × × × × ×

BARNSTAPLE JS. ROLLE QUAY
211 M. 49 CHS. SIDING
212 M. 41 CHS.

GEORGEHAM X'ING
217 M. 68 CHS.

VELLATOR X'ING
217 M. 2 CHS.

COMMERCIAL ROAD X'ING
212 M. 9 CHS.

651 294 449 243 230 660 1456 868 LEVEL 474 510 1026 504 314 628 772 420 750 266 125 246 1760 366 132 236 74 96

MILES 212 213 214 215 216 217 218

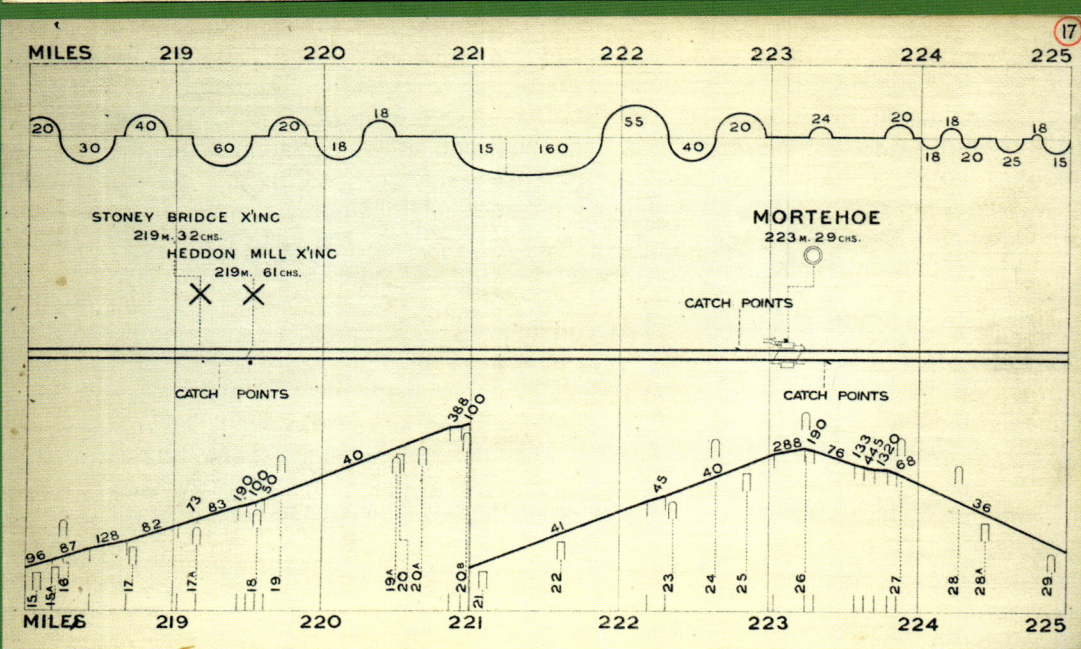

MILES 219 220 221 222 223 224 225

20 40 20 18 55 20 24 20 18 18
30 60 20 18 15 160 40 18 20 25 15

STONEY BRIDGE X'ING
219 M. 32 CHS.

HEDDON MILL X'ING
219 M. 61 CHS.

MORTEHOE
223 M. 29 CHS.

CATCH POINTS

× ×

CATCH POINTS

368 100 CATCH POINTS

288 190 76 133 439 130 68

40 45 40 36

96 87 128 82 73 83 190 150 41 208

MILES 219 220 221 222 223 224 225

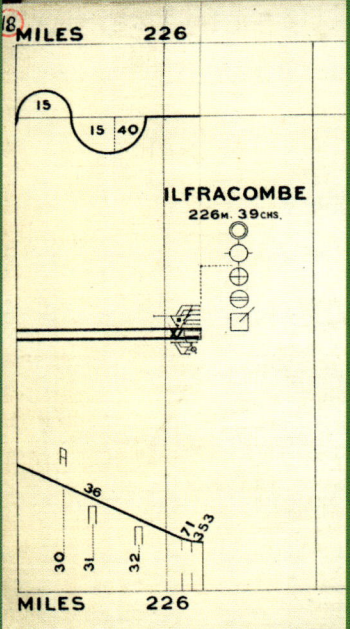

MILES 226

15
15 40

ILFRACOMBE
226 M. 39 CHS.

36
71 353

30 31 32

MILES 226

Extract from Book 8 of the LSWR 'Diagram of System' maps - undated.

refused. Soon after BR wrote to the Minister advising that in excess of £300,000 was needed to effect replacements of the four bridges involved but like a spoil child not prepared to accept defeat, repeated their assertion it was a waste of money. The original target for closure of the Southern lines had been August 1965 whilst curiously there is no mention of how traffic from Meldon might have been handed.

The 'Chippenham' papers appear to have their first reference in 1962 when mention is made of staffing levels on the line even so only two locations are mentioned, Mortehoe where there were four staff and Ilfracombe where there 16, the latter number including four who were supernumerary and only engaged during the summer. Usually, we might expect such figures to refer to Traffic Department personnel only but jumping ahead to what is quoted when eventual closure does occur it could in fact refer to staff in all departments. The two stations mentioned were also just two out of a large number of former Southern locations so right at the outset we might ask was this a regular management exercise or one conducted solely on lines considered at risk? At the same time the comment was made that often the staff are required to work overtime during the summer and the service should be adjusted to remove this need.

With the Western Region saddled with what were realistically a number of loss-making lines within its territory it cannot have been a surprise to realise they would do everything they could to reduce these liabilities. Aside from needing to be seen to make a profit, reducing loss was essential in order to secure investment for the future. There was also the question of replacing steam with more modern traction, each of the regions receiving the same blunt message 'Reduce overheads to release funding for new projects (traction)'. This then was the era when managers might receive a bonus for the savings they achieved, seemingly regardless of how this was achieved, and that of course included what were considered to be loss-making services.

The years from 1963 on were a time when various surveys were carried out 'bean-counting' in effect or more accurately 'passenger headcounts'. The first of these located was for the week commencing 10 November and involved all Up (to Barnstaple) and down (to Ilfracombe) services. The numbers gave a foretaste why closure was being considered, an average headcount of just 199 passengers daily for Down trains with the mid-day services the busiest. Up trains carried just a few more; 210 persons. On Saturday the figure was broadly similar but on Sunday

there was a drop to 77 people on the Down trains and 131 on the Up.

Before the reader starts jumping up and down to protest this was a typical census taken during the winter months let us add that the exercise was repeated in the week commencing 5 July 1964. Now the figures were different and whilst confirming Dr Beeching's assertion that many lines were 'seasonal' the figures do make for an interesting comparison. Weekdays the numbers were in effect double with 429 passengers spread over 11 services. On Saturday it was 1,801 over 17 trains the 1030 ex Ilfracombe having 343 passengers. The Sunday workings had 200 persons in six trains.

Figures but this time '£s' come into play again towards the end of 1964 in a report prepared by the 'Special Financial Services Section' at Reading for the attention of Mr C P Hopkins the Divisional Manager at Plymouth. The report stating, 'Re your letter of October, for the 13-week period ended 3 October the figures are.

Through Forwarded			
	Local Receipts	Gross Earnings	Branch Receipts
Ordinary Passenger.*	£5,000	£56,800	£3,400
Seasons		£50	£10
Special trips		£50	
	£5,350	£56,900	£3,410
* Includes Government Warrants. Total contributory revenue of £157,150			
Through Received			
	Branch Earnings	Gross Receipts	Branch Earnings
Ordinary passenger.*	£7,050	£172,300	£15,450
Seasons		£400	£350
Special trips	£40	£300	£40
	£7,090	£173,000	£15,850
Receipts in respect of Holiday Runabout Tickets issues on the branch during the period amounted to £830.			

A copy was also sent to Paddington whilst it is interesting to note the numbers appear to make no separate reference to goods receipts.

Mr Hopkins or 'A N Other' on 29 December 1964 were quick to make what were brutal recommendations to Paddington. These were to propose to retain only passenger services for which a subsidy of £7,500 would be required per annum and also to single the line. (This is the lowest subsidy figure ever quoted. It was around this time that were also only two real options for rail services; profit or closure. In the case of recommended closure this could be negated if the minister of transport would agree to a subsidy – perhaps it is of everlasting shame the WR did not attempt to approach Government for what was a relatively small figure, who knows it could well have succeeded and starved off the immediate threat of closure. We can only speculate why this was not done, perhaps because the WR still retained an avowed intention to close everything west of Cowley bridge.) Other proposed alterations were to retain the platform only at Barnstaple Town, at Wrafton to retain the down platform only, at Braunton to retain one platform only and the road frontage at the level crossing the station buildings not required, at Mortehoe to retain the up platform only and again the station buildings were not required and finally at Ilfracombe where just the Barnstaple end of platform would be retained together with the former goods office. As a comparison should the route be closed the land value was set at £2,000.

More head counts, costs and this time staffing costs followed soon after, this time week commencing 21 February 1965:

	Monday - Friday		Saturday	
	Joined	Alighted	Joined	Alighted
Barnstaple Junction	128	167	158	186
Barnstaple Town	175	142	198	183
Wrafton	14	20	15	11
Braunton	75	68	50	56
Mortehoe & Woolacombe	26	29	30	28
Ilfracombe	67	59	92	79
Totals	485	485	543	543

Staff numbers / costs were given but only for signalmen or men engaged in such work for part of their daily study:

Ilfracombe	2 Signalmen Class 3
Mortehoe	2 Signalmen Class 4
Braunton	2 Signlamen Class 3
Crossing Keppers	4. Total £2,194. 4s
Crossing Keepers	2. £1,097 2s. (Rest day relief)
Warfton Leading Porter / Signalman	2. £1,229 8s.
Crossing Keeper	1. £548. 6s
Barnstaple Town	2 x Signalmen Class 4 2 x Signalmen Class 4 (Pottington) 1 x Signalman Class 2. (RTR.)

Other staff numbers probably from the same time were:

Ilfracombe	1 x Class 3 Clerk 3 x Class 4 Clerk#1 x Class 3 Station inspector 2 x Shunters - Passenger 2 x Porters 1 x Leading porter 1 x Senior porter 2 x Piorters (Summer Only) 1 x Office Cleaner (Female) part time
Mortehoe	2 x Leading Porters
Braunton	2 x Grade 2 Station Foreman 1 x Grade 3 Clerk 2 x Porter
Wrafton	Nil
Barnstaple Town	1 x Class 3 Clerk 2 x Leading Porter

Interestingly there is no mention of anyone in the role of Station Master anywhere on the line.

Jumping ahead slightly the BR files show an annual comparison of revenue at stations on the line for the years 1965 – 1967

Revenue	1965	1966	1967
Barnstaple Town	£1,067	£13,600	£10,167
Warafton and Braunton combined	£10,486	£11,159	£584
Mortehoe & Woolacombe	£4,473	£8,924	£2,890
Ilfracombe	£23,566	£20,082	£14,791

We may assume these are as before based on tickets issued on the line and not returns which was as before disadvantaging the route. The one anomaly has to be 1965 figure for Barnstaple Town; we can only assume a clerical error.

The sphere of activity now moves to Exeter St Davids and where on 5 August 1965 an unknown official in writing to Mr Hopkins at Plymouth states that in (another) census taken week commencing 11 July, numbers show that by withdrawing the Ilfracombe passenger service there would be a net saving of £47,000 exclusive of £16,700 renewals due in next five years. It was admitted these figures were based on the present full staffing and the existing double line railway. On the opposite side, the Revenue Accountant (?) estimates that the line contributes £201,500 annually and if closed a minimum of 10% of this contributory total would be lost. This 10% number was once again BR with their 'pie in the sky' attitude that existing passengers would then make their own way to the next available railhead, in reality we know this rarely happened. (A more realistic appraisal of this type of situation came from Former Divisional Manager, Mr Pattison who considered a loss of 33.3% of contributory revenue would be more accurate and

Driver's eye view approaching the terminus at Ilfracombe. Mention is made in the text of the different 'number crunching' exercises made one of which showed a regular 13 commuters and 5 children on weekdays. No mention of hardship or difficulties affecting these groups appears to have been referred to at the TUCC enquiry. *Transport Treasury*

even this was probably an under-estimate.) The memorandum concluded that it is believed the saving could be reduced to £27,000 and that it might be possible to operate the railway for £37,000 annually. 'This would appear to enable this section of line to break even on a minimum staffing basis.'

These numbers were not enough to persuade the Western Region to capitulate, and the next stage was an investigation into comparative costs of rail – v- road travel between Barnstaple and Ilfracombe as well as capacity issues at Barnstaple. So far as alternative buses were concerned the difference in fares between road and rail was pennies (old 'd'), on certain journeys the rail fare cheaper and on others it was the bus. Not surprisingly where the train scored was in regard to travel time, the train being quicker in every case. Rail travel to both Mortehoe and Ilfracombe 10 minutes quicker by rail and considerably longer on occasions due to local road congestion.

The exercise was repeated the following year, perhaps BR had been hoping to find bus fares were now considerably cheaper (did they raise rail fares by an unreasonable amount in the interim?) but the comparison was the same. Indeed, there was now an additional issue for the railway had to admit, Barnstaple is 'bursting at seams' following concentration of traffic from the Bideford line and passengers awaiting buses. The writer added it was impractical to deal with more from the Ilfracombe line.

BR however would not leave the subject of an alternative bus service alone, no doubt wishing to make a case that the present bus service could accommodate the existing railway traffic. But the local bus company, Southern National, were not going to allow themselves to be made the scapegoat and responded that their own July 1965 survey which revealed they would need a minimum of six 68 seat double deck vehicles in continuous operation between Ilfracombe – Mortehoe – Barnstaple from 0615 to 2200 which tends to imply their own estimate was that there were more passengers for Ilfracombe than the railway as admitting. Their available resources also meant they could supply only three buses the reason being shortage of staff at the Barnstaple and Ilfracombe bus depots in summer months when demand for seasonal labour is high. Their conclusion being that it would be extremely difficult to meet these additional requirements. No mention is made of the luggage that would have accompanied holiday passengers.

Confirmed in writing from within the Chippenham files was the hope BR could rid themselves of the Ilfracombe service in 1965 but with such comments from Southern National it was realised this would not be possible – for now at least. So, the railway had tried and for the present failed, they would not give up.

Plymouth (and Paddington who were of course involved) were not to be beaten and now tried a different tact based on infrastructure expenditure. This was to calculate the estimated cost of repairs and maintenance resulting in a memo from the Chief Civil Engineer at Paddington dated 5 December 1966.

Five yearly prognosis	
Ifracombe	Relay part crossover in up line with new material £1,550
Mortehoe	Relay double catchpoint in down line with new material £1,050
Years 3-5 Pottington to Wrafton	Re-sleeper down line £2,850
Braunton to Mortehoe	Relay up line with new materials £43,475
Repairs to Pottington swing bridge at 120m 12ch	Estimate of £35
Total	£49,275

Contrary to previous folklore there is no mention of repairs to the bridge over the River Taw at Barnstaple.

All this time 'closure by stealth' was either taking place or in the planning stage. Freight services had been or were withdrawn by 1966 and the following year, 1967, plans were being drawn up for singling to take place leaving the barest minimum service - until the inevitable closure occurred. The one hope was for a subsidy the best chance of this being if closure was proposed giving the Minister of Transport the option.

Singling the railway was therefore effected from 17 December 1967. The existing up line utilised from Pottington to Braunton, the down line from here to Mortehoe and finally the up line again to Ilfracombe. At Ilfracombe itself the layout was reduced to a run-round loop and siding. Signalling and with it line occupancy was similarly pared back with only Barnstaple Town remaining as a block post. From Barnstaple Junction the whole of the remaining route to Ilfracombe was operated on the 'one engine in steam' principal' with a wooden train staff carried by the engine attacked to which was an Annett's key to unlock ground frames at Ilfracombe. Whilst maintenance of the permanent way will have been reduced and several signalmen's posts saved there was an increase in the number of crossing keepers required; the various level crossings still protected by stop and distant signals.

The effect of the singling was to also restrict the number of trains that could use the line. Travel time between Barnstaple and Ilfracombe (and vice-versa) was in the order of 40 minutes each way and with no

means of 'locking a train in' at Ilfracombe this meant an approximate headway of at least two hours between trains. Even so there were still some through workings from Paddington during the summer although with the closure of the Taunton – Barnstaple line in 1966 all such trains ran via Exeter St Davids where a reversal was required. The subject of through trains created much correspondence with a number of attempts made to curtail these at an early stage.

From 30 September 1968 a Conductor / Guard scheme was introduced on the Ilfracombe and Barnstaple lines meaning tickets were issued on the train rather than from a station booking office. Whilst this might have worked in winter it was no consolation at Ilfracombe in summer where the station foreman had no choice but to turn passengers away from the one remaining through service to Paddington should they have failed to obtain a compulsory seat reservation.

Conductor operation was considered to save £4,500 annual with an estimated additional saving of £5,000 possible if the various level crossing were to be converted to automatic half barrier working although no

ALTERED FACILITIES AT STATIONS ON THE EXETER-OKEHAMPTON AND EXETER-ILFRACOMBE LINES

From Monday, 30th September 1968, if you join a train at Ilfracombe, Okehampton, Newton St. Cyres, or at any station in between you should obtain your ticket from the Guard except at Barnstaple Junction where the existing ticket issue facilities will remain. The Guards will be capable of issuing tickets to all destinations within the area bounded by Ilfracombe, Okehampton, Paignton and Plymouth, also Exmouth and principal stations to Bristol and London. If you are travelling to a destination to which the Guard is not capable of issuing a ticket you should take a ticket to Exeter and re-book there. Apart from Cheap Day Returns which are available between certain stations as advertised, the Guard will only issue single tickets. Please help the Guard by stating your destination, type of ticket required, and tendering the exact fare.

Season tickets should be obtained from Exeter St. Davids, Exeter Central, Okehampton, or Barnstaple Junction stations by personal application to the Booking Office or postal application to the Area Manager.

For train information please ring EXETER 77277/8/9, BARNSTAPLE 3162, or OKEHAMPTON 2633.

Details of the Cheap Day Fares are shown in this leaflet, also the train service operative until 4th May 1969.

Facilities for despatching parcels from or sending them to be called for at Newton St. Cyres, Crediton, Yeoford, Bow, North Tawton, Copplestone, Lapford, Eggesford, Kings Nympton, Umberleigh, Wrafton, and Mortehoe & Woolacombe stations will no longer exist. At Barnstaple Town station parcels may no longer be despatched but received traffic will be dealt with for collection by the public up to 20.00 hours on weekdays. Facilities for despatching or receiving parcels continue to exist at Ilfracombe, Braunton, Barnstaple Junction, Okehampton, and Exeter St. David's stations. Enquiries regarding the collection and delivery of parcels traffic in the area should be made to the Freight Depot Manager at Barnstaple (Telephone 4171) or Exeter (Telephone 72251).

1

Western Region DMU service - undated. Full dieselisation of services on the line took place from September 1964. For a time there remained through workings, but by DMU to Ilfracombe from Salisbury and Exmouth better than nothing but still a far cry from the former 'ACE' restaurant car service. Post war the peak service for passenger comfort was in the Pullman cars of the Ilfracombe portion of the 'Devon Belle' but this had of course operated for only a few years. The peak passenger flow was said to have been in the summer of 1958 after which there was a gradual decline. In many respects the railways might even have been considered a victim of their own success, able to move the masses but the masses were beginning to want their own, private, alternative. The last through working to Paddington was on 26 September 1970. *Transport Treasury*

cost is given for this. A further saving of £10,000 could be made if all locomotive hauled services were withdrawn ('top and tail' working was not mentioned). This last number seems strange as surely the only alteration would have been the removal of the run round loop, siding and ground frame at Ilfracombe; perhaps reduced track maintenance for DMU only operation was a factor.

Another suggestion was to adopt Barnstaple Town as the principal station for the town, this owing to its proximity to local bus services and that Barnstaple was even being considered for development for London overspill.

Closure proposals were again put forward in 1968 based on figures which whilst admitting the number of persons who might join an Ilfracombe train at Barnstaple in summer could be 300, in the winter that figure could be as low as one. BR's case was based on economics and with two figures quoted for renewals. The first was a perhaps reasonable £12,210 which in the next document is suddenly inflated to a whopping £132,310! Even so it must be admitted there were different loss figures shown in almost every file.

This large variation had not gone unnoticed internally either for on 20 February 1968 there is a memo from the Principal Costings Officer Paddington to BRB entitled 'Track costs/expenses Barnstaple – Ilfracombe.' 'Referring to my earlier conversation with you I have continued to probe the reasons behind the different figures for track costs/expenses which we are producing and using for passenger grant URS/TUCC submissions. This examination reveals a situation which appears extremely hard to justify; press publicity might do us serious harm.'

We hesitate to suggest but is this the 'smoking gun' discovered at last?

Undeterred, closure proposals were announced and which brought forward a raft of objections and resulted in a public enquiry held by the Transport Users Consultative Committee at the Palace Theatre Ilfracombe at 1045 on 9 October 1968 Chaired by Lt Col J K MacFarlan OBE. The TUCC had received no less than 371 private objections to closure plus 38 objections from organisations, including the local Congregation Church in Ilfracombe, the whole amounting to 74 typed pages. At the enquiry there were 21 speakers against closure plus a representative (not named) from BR. It had been anticipated that the Traffic Manager from Southern National would be attending but shortly before the date he cancelled citing 'Staff Shortages'. His note to decline added, the Traffic Manager '…sees no useful purpose…' in attending the enquiry (where no doubt he would face awkward questions. (He would send a 'minion' instead!)

At the enquiry the BR representative stated, 'Mr. Chairman, members of the T.U.C.C., ladies and gentlemen, I thank you on behalf of the B.R.B. for giving me the opportunity to make an opening statement on this occasion. I hope that what I have to say will set the scene and provide a sound background for hearing the various objections to the Railway Board's proposals.

'You are, of course, aware that the Railways are under a statutory obligation to pay their way and it should be made clear that the amendments to railway deficit grant financing embodied in the Transport Bill currently before Parliament do nothing to relieve this commitment. Indeed, the law in this respect is to be strengthened by relieving the Railways of the burden of many of the known loss-making activities such as branch lines, bridges, police, etc. and requiring money to finance future losses to be borrowed on the open market at current bank interest rates. (There was already in existence a £140,000 pa grant / subsidy to maintain services between Exeter and Barnstaple. This was not referred to at the hearing but from correspondence we learn BR were concerned at the time as to if this would continue.)

'In common with many other branch line services in the West Country, the Barnstaple to Ilfracombe line loses money and is one of those services which will require long term financial assistance if it is to be retained. The Minister of Transport alone will decide whether the social and economic benefit to be obtained from the maintenance of a particular service is sufficient to justify the cost of retaining it. The purpose of this meeting is to assess the degree of social hardship which might be involved in the event of closure.

'The possible closure of the Barnstaple/Ilfracombe line was foreshadowed as long ago as 1963 upon the publication of the Reshaping Report, more commonly known as the Beeching Report, but until fairly recently it was left to the broad discussion of the Railways Board to decide whether or not to publish their intentions to close a line. One of the reasons for the Railways Board's reluctance to publish the Barnstaple/Ilfracombe line for closure has been the knowledge that there was some difficulty on behalf of the bus operators to provide an adequate alternative service. This was one of the particular factors which had to be taken into account in earlier closure cases, and the ability of the bus company adequately to cover the commitment was, in fact, a requirement which had to be demonstrated.

'The Minister is aware of the stated deficiency in this particular case (and I expect the bus company's representative will comment upon it) but the Minister has nevertheless instructed the Railways Board to formally publish an intention to close the line in order that the alternative road service feature may be publicly aired. In fact, it is true to say that in order that the line may be properly considered for keeping open, we are compelled firstly to publish intention to close it.
'In the cases of some of the West Country branch lines, the Minister has already decided to pay a subsidy without a formal public enquiry; in others such as these a decision is awaited, but in this case he requires more information which it is hoped this hearing will provide.

Thank you again for affording me the opportunity of making this statement.'

When the railway's representative was questioned on the subject of road congestion, he neatly side-stepped by stating, '...I regret I am unable to comment on matters of road congestion, this should be referred to the Ministry of Transport.'

We are not told what other statements and by whom were made at the enquiry but the below are a small sample of some of the written objections;

Separate letter of 1968, from Trinity House Lighthouse Service 'It is understood that the railway line between Barnstaple and Ilfracombe may be closed......the railway is used in the supply of stores to essential seamarks on the North Devon coast an example being the Bull Point Lighthouse which is served from Mortehoe.'

(The various objector's letters were numbered.) Some samples follow commencing with No 34A from Mr Britton. 'The Railway is the only realisable means of transport in this area; bus services are inadequate and unreliable. They offer no facilities whatsoever for the carriage of prams and pushchairs and make charges for parcels and packages accompanied by the passenger. The fares are dearer than the rail fare, and on some journeys the fare by bus will be 30% more than the rail and bus journey combined. Connecting bus services at Mullacott Cross are hopeless, and there are no facilities for passengers who will be waiting here for anything from 15 to 45 minutes for a connecting bus. Mulacott Cross being noted as a very bleak spot and the bus company have refused to provide any covered bus shelter at this point. The withdrawal of the rail service will affect my own livelihood and the traders and hoteliers in the district and have a very adverse affect generally. Also, the amount of traffic that will be driven on to a very poor road, of which there are numerous accident black spots. The bus service could not possibly cope with the volume of summer passengers in the resorts of Saunton, Georgeham, Braunton, Woolacombe, Mortehoe and Combe Martin.'

35A Miss V Tucker. 'I travel to Ilfracombe frequently where my mother owns some property. The journey from any place on the Exeter side of Barnstaple will be impossible as so much time will be wasted co-ordinating trains and buses (which are very slow). My mother is elderly with no alternative means of transport who travels to Ilfracombe several times a week and will be seriously inconvenienced. She is one of many such people. Those living in the country villages along the route often have no other link with the towns of Devon and those people often do not realise that a complaint may be made and their case heard. I would like to draw the attention of the Committee to people such as this

Interior view of an 'empty' DMU between Ilfracombe and Mortehoe. By 1970 the daily service had dropped to just five trains each way compared to 11 in 1963. This might even be said to be closure by stealth but then to be fair to the railway it was also a 'Catch 22' situation. If few people travel then they reduce the service but that then attracts even less patronage whilst the overheads remain the same. *Transport Treasury*

who will ultimately suffer the most. The Devon roads are most unsuitable for the extra traffic that a closure of a railway line will inevitably bring. The holiday traffic of Croyde, Woolacombe and Ilfracombe will suffer very much. We have already been seriously inconvenienced by the closure of the Bideford line, and my mother was very ill after waiting for a bus in the rain.'

36A Mr D Sanders. 'I very much regret that so little thought for the elderly is given when closing these lines. I live on the 'late' Bude line, Okehampton to Bude, and it has compelled me to walk on roads which have now become really dangerous for walkers, owing to the large unsafe lorries and their loads which now have to be carried on such narrow roads as the B3218 with all its bends and hills. I have taken going to Ilfracombe instead of Bude and it will deprive me of my chance of pleasure in my old age (I am 74) if this line is closed. I use the line three times between June 10 and 17 on my 7-day Runabout ticket. I have 5 or 6 each season at least instead of a set holiday, as well as my relations who have them when visiting me here. May I add that buses do not take the place of trains for many people cannot for one reason or another use them.'

41A Mrs E O Abbott. 'The withdrawal of trains on the Ilfracombe line would cause hardship and inconvenience when travelling with young children and prams.'

42A Miss A. Abbott. 'The delay caused by congested roads when travelling to Barnstaple would cause great inconvenience and possibly missed trains.'

43A Mr K. G Abbott. 'The increase in fare when travelling to Barnstaple would cause hardship.'

44A Thelma C Northcott. 'This service is vital to me as it would be quite impossible for me to get to my work in any other way. To go from Saunton to Mortehoe station (place of my work) would take me nearly two hours and three buses.'

45A Mrs E M Soloman. 'It will cause me as an old age pensioner quite a lot of hardship as I cannot travel by bus at all owing to ill health therefore, I should not be able to see my grandchildren who live down in Devon.' Un-numbered Mrs. Hazel Shutts. 'I am a housewife with two very young children - the train is much more convenient and comfortable than the local bus service. Setting aside personal reasons I feel that as Ilfracombe is a development area needs all the communication systems possible. Lack of a rail link between Ilfracombe and Barnstaple will surely deter many industrialists, not to mention those holiday visitors who do not own a car.'

56A Mr P W Young. 'As my wife and self are frequent railway travellers between Ilfracombe and Salisbury and using a non-smoking compartment, as this suits me better being a great sufferer from asthma. If the line is closed, we will have to use the buses and if the downstairs were full and we have to go upstairs where smoking is allowed it would be detrimental to my health.

57A Mr. P.H. Lason, Hon. Treasurer (acting Secretary), Combe Martin & District Chamber of Trade. 'I have been instructed by the General Committee to protest in the strongest possible terms at the proposed closure of the Barnstaple, Ilfracombe Railway line. We have as a body circulated forms that have been completed by many members of our organisation proving that their concern is great enough to warrant fighting this stupid and unwarranted decision of closing one of our main lifelines of transport. These forms were deposited with our main Federation and the main objection being the very obvious reason, the complete inadequacy of our roads to cope with the ever-increasing heavy traffic.'

58A Mr S A Hatchley. 'I travel by train to Barnstaple once or twice every week on business and if the railway closes, I shall have to travel by bus which means extra money. I am only on old age pension so it will be harder for me to pay extra money it is a job to pay your way now without more expense.'

59A Mr G T Pearson. 'I use rail whenever possible as bus travel upsets my wife and travel frequently to Exeter and South Devon. There is no bus service from this area to Exeter. Bus services between Braunton and Barnstaple not adequate for train connections there. Braunton station is convenient for passengers living in this rapidly expanding area and with a more suitable train service would justify its existence. The present timetable between here and Exeter appears to have been arranged with the intention of driving passengers away and discrediting the line, so as to make the case for closure convincing. The last train from Exeter to North Devon is much too early for distant travellers to catch them. I feel that any move to close lines, especially to an area such as ours with its holiday industry and its status as a development area is very short sighted of the Board. In a matter of only a few years people are going to be keeping as far away from the roads as is possible, surely this is a large part of British Rail's

advertising campaign and a decision to close the line would be inconsistent with your policy. I fee that in this letter I speak for all students, young people and others who use or will want to use the Railway service and have not lodged a complaint.
'

60A L. J. Bayliss. 'It would be virtually impossible for a shuttle coach or bus service to convey holidaymakers such as myself between Barnstaple and Ilfracombe due to the extensive number of passengers involved. The withdrawal of daily through services from London has been a great enough inconvenience, so the suggestion that passengers from Surrey and Hampshire etc., should change twice i.e., at Exeter and Barnstaple is out of the question if Ilfracombe and Woolacombe are to continue to be resorts for other than car travellers.
'

There was also the almost pathetic case noted of a retired railwayman whose objection to closure was based on the fact he would not receive concessionary travel on the bus when travelling to Barnstaple to shop.

At the same time BR did submit a subsidy application to the Ministry in August 1968 but with a disparity in the figures applied for in the Grant and the actual BR Unremunerative Railway Service.

	Grant Submission	URS Submission
	Costs p.a.	'Expences' p.a.
Earthworks, drains, fences	£2,320	£3,300
Bridges, tunnels, retaining walls etc.	£9,650	£830
Permanent way	£19,882	£12,730
Totals	£31,852	£16,860
Administration	£4,778	
Total	£36,620	£16,860
Interest	£6,240	

Sometime in the same year, 1968, the focus of BR activity moves from Plymouth to Bristol and where a summary of the current situation was made available to the Divisional Manager Bristol. This This confirmed the railway as currently the subject of an application for closure based on it being an Unremunerative Railway Service. The current situation reported as; at that time there was a small nucleus of shoppers and visitors throughout the year with the railway busy on Summer Saturday; all trains are through services to Exeter. It was noted the Ilfracombe holiday season also extends beyond the railways' own Summer timetable. All stations on the line are relatively well used excepting Wrafton (for RAF Chivenor) where traffic is virtually confined to the adjacent RAF base. Originating parcels traffic provides revenue of around £3K annually. Passenger receipts at Ilfracombe are declining, cheap day bookings 40% down in June 1968 compared with previous 12 months. This indicates we are loosing the local, regular and short distance passengers perhaps because of uncertainty of the future of the line rather than the main line holiday travellers. Here are no services on Sundays.

Jumping ahead slightly, the following year loading figures for the one remaining through train from Paddington were also given but with no comparison as to previous years:

0810 Paddington - Ilfracombe. Loadings 1969	
14 June	347
21 June	258
28 June	321
5 July	297
12 July	233
19 July	450
26 July	450
2 August	450
9 August	245
16 August	468
23 August	378
30 August	264
6 September	131

Reverting back to 1968 and on 25 July a letter was sent from solicitors acting on behalf of Sir Robert Williams drawing attention to a conveyance of 1874 in which at least two trains daily shall stop at Wrafton for his use and also by request for any other residential member of his family. This was a not-uncommon covenant from the time railways were in the phase of construction and was in effect often a sop to the local landowner allowing the railway to be built across his land. The response from BR was to quote Section 36 of the British Railways Act of 1963 which effectively quashed such requirements. In some circumstances financial compensation was then paid to the landowner although we do not know if it applied in this case.

In late 1968 the TUCC reported on their enquiry, their findings sent to the Minister on 4 November. The conclusions were, 'The Committee are of the opinion that implementation of this proposal (for closure) would cause hardship in varying degrees during the summer holiday season to users throughout the area of the line and to those who use it for travel further afield. Hardship would also occur at all times to people using Mortehoe and Woolacombe station for journeys to and

from Barnstaple and beyond.

'The committee are also of the view that thousands of holiday makers now visiting Ilfracombe and North Devon would not or could not continue to do so if rail access were withdrawn. This would have a marked effect on the economy and trade of the area and hardship would result to many who are dependent on the holiday industry for their livelihood.

'The committee, like the British Railways Board, are unable to suggest any means of alleviating hardship if the trains are taken of. The committee were unanimous is reaching these conclusions.

'It is desired to call attention to public concern about the continued uncertainty of the line's future and with the hope expressed at this hearing that a long period of doubt would soon end and so give the transport user a new sense of stability.'

It could not have been put more succinctly; the Ilfracombe line should remain.

There was also a final census of traffic, this time parcels, and dating from 9 October 1968.

Barnstaple Town: received approx. 10 x 150 chips of mushrooms ex Braunton per week. Forwarded approx. 20 packages per week of general parcels. 12 packages of pram spares and pathological specimens per week.

Wrafton: received approx. 20 packages per week from the RAF. Forwarded approx. 4 packages per week from the RAF, other traffic here was negligible.

Braunton: Received approx. 50-60 general packages 'To be Called For' per week. Forwarded during the past 12 months some 65,000 packages of flowers, mushrooms and rhubarb. In May 1968 3,000 packages of flowers sent to about 10 destinations and 2,000 chips of mushrooms sent mainly to Plymouth and Barnstaple; the other main destination was London. Approx 10 ordinary parcels forwarded per week. During past 12 months 125 items of delivered luggage were despatched, Newspapers, 25-30 bundles received each day.

Mortehoe: Received approx. 7-8 parcels per week. Forwarded approximately 2 per week. 'Passenger Luggage n Advance' dealt with, approximately 5 items per week during the summer months.

Ilfracombe: received approximately 35 parcels per week on average. Forwarded approximately 25 per

DMU service at Ilfracombe - with few passengers. *Transport Treasury*

week average. . 'Passenger Luggage in Advance' approx. 14 per week average over a 9 monthly period. 10cwt fruit per week sent out in the holiday season plus 6 boxes fish per week. Newspaper traffic 40 bales received per day which expands to 110 bales per day in holiday season.

All of this was lost when the line was singled and DMU operation took over. Around the same time there are details of redundancies. The total number of staff affected being 58 including five drivers and two second men, three goods guards, 11 gangers / length men, plus 37 traffic department staff from supervisory grades to clerks, signalman, porters, crossing keepers and a female office cleaner. Just eight of the staff were redeployed.

It took more than a year for the closure decision to be made, exactly the fear the TUCC had cautioned over. When the decision was made it was that the Minister of Transport Fred Mullay would consent for closure, this being made public on 31 December 1969.

Figures quoted to justify the claim were revenue of £13,300 against expenses of £93,300. Broken down the £93,300 referred to track and signalling at £56,600, movement cots of £13,500 and £23,400 for 'terminals'. Nicholas and Reeve make the point how this last figure could relate to the limited track remaining at Ilfracombe is difficult to imagine.

The Western Region were quick to arrange for the necessary posters to be printed, 50 ordered from the Plymouth printing house of Latimer Trend (a supplementary note made a request for one further copy soon after). The posters were distributed in the area Bristol to Plymouth including local stations.

Politics now rear their face in the form of the charismatic local MP for North Devon Mr Jeremy Thorpe – those of a certain age will be aware why the present writer uses the term 'charismatic'. There appears to have been no mention of Mr Thorpe prior to this point but on 12 February 1970 a newspaper cutting from 'North Devon Journal and Herald' reads, 'Liberal leader Jeremy Thorpe fighting the proposed closure of the Barnstaple

– Ilfracombe line and is demanding a slow-down from British Rail. He has claimed British Rail are using bogus figures to support a closure…. .'

Just over a week later on 20 February the 'Western Morning News' stated that Mr Thorpe has asked the Parliamentary Ombudsman to investigate the case on the grounds that the decision of the Ministry of Transport at the end of last year was based on figures 'which to say the least are highly questionable'. Mr Thorpe gave the example of track and signalling expenses for 1969 put at £57,000 with his own estimate of £10,000. No doubt with an eye towards his constituents, Thorpe did not mince his words and in a way other lesser mortals might not have been able to stated,' One begins to wonder (this in a letter to the then BR Chairman Sir Henry Johnson), what other unknown expenses have British Railways Board added on and what indirect or overhead expenses forming part of the operating costs

of British Railways as a whole have been pushed on to this 15 mile line. I suspect that a very large proportion of main line costs have been debited to this branch line. On that basis almost any branch line could be rendered uneconomic.' He had a point as well, that being the accountant's methods of not crediting a branch line with its value in return tickets as was mentioned at the start of this piece.

The paper added that last weekend Mr Thorpe had taken his own private census of traffic. At Ilfracombe 969 arrived and 767 left. At Braunton 162 left and 206 arrived. At Wrafton 99 left and 80 arrived, at Barnstaple Town, 109 left and 103 arrived. His average was 50-60 persons per train. We are not told of the date this particular census was taken.

Despite Mr Thorpe's protestations there was now no way back with closure announced to take effect

There were few scenes sadder than those of stations serving seaside resorts which had been become largely deserted. Partly this was due to changing travel habits and holidaymakers in recently acquired cars but in the case of Ilfracombe it was also a case a rail service that was unsuited to the masses. In its last days the spacious hilltop station hardly resembles what had once been a bustling terminus and the destination of two named trains. In this image the threat of imminent abandonment was palpable. The large derelict structure to the right had been the goods shed, the engine shed and signal box behind the photographer similarly abandoned. In this view from Wednesday 26 August 1970 we should be in peak season but instead the site is deserted.

after the last trains on Saturday 3 October 1970. Six weeks before what is almost certainly an enthusiast working in the Marketing and Sales group Bristol sent a note around reminding various departments that the passenger train service will soon be withdrawn and, 'How about a few special blackboard notices to draw in the day trippers and rail enthusiasts whilst they still have the chance to sample this line?' We do not know if anything resulted. What we do know is that the final train, an 8-car DMU, ran at 1955 on Saturday 3 October with 500 passengers.

There matters might have ended except for a preservation attempt by the newly founded 'North Devon Railway Preservation'. A share issued was organised with £650, 000 shares offered at 10p each and £100,000 in £1 shares. British Railways had valued the land, buildings and remaining track at £300,000. The NDRP countered this with an offer of just £50,000. There is a suggestion, not completely clear, that the Ilfracombe Urban District Council had expressed an interest in purchase with a reply from BR that 'they must be quick'. The NDRP had the moral support at elast of another charismatic player of the time, Sir Gerald Nabarro who had been involevd in the early days of the Severn Valley Railway. Whatever, the NDRP company were unable to raise funds and the deadline to make interest payments to BR to retain the line in situ passed. The company was disolved in 1973.

Further detail of this failed attempt is really out of place in this text and for more information the reader is again referred to Nicholas and Reeve page 277.

In consideration of the attempt to save the line the rails had been left in place and on Wednesday 26 February 1975 one final train was seen to cross the River Taw, pass through the various level crossings, the stations and climb the bank at Mortehoe to reach Ilfracombe. This was a BR 'special' hauled by a Class 25 consisting of an inspection coach and with the hardcode '1Z01'. The purpose was to survey the line and buildings, the special having to stop at every crossing to open and shut the crossing gates. Figures indicate the necessary 'fettling' to allow this last train to run had cost £330.

It also really was the last train as soon after contractors moved in to lift the track, the latter taken into storage for re-use. The cost of recovery put at £81,670.

Half a century later the senior players in this debacle will all have passed on. The managers who by closing railways no doubt felt they were carrying out their duty, the passengers who were forced to transfer to buses often without the ability to carry the prams, pushchairs and the amount of luggage required. It was closures

like these that led to the speeding up of car ownership, inevitable though that was.

We might wonder if Mr thorpe was right when he said overheads were being placed on the line to make it appear loss making. There does not appear to have been any reply to his questions to the Ministry on this point; perhaps any reply was too 'hot' politically.

BR's method of accounting has a lot to answer for as well. The method then in use meant only the strongest survived but also the figures could be manipulated to prove almost any line was at risk. It was indeed only by good fortune that Barnstaple remained - perhaps that was simply one closure too far. The objectors points about buses being unable to cope were well founded, the TUCC agreed. It was the same argument, along with road congestion that was used for many railway closures. What is not clear from the Ilfracombe papers is if the railway had to subsidise the bus service after rail closure. This certainly happened elsewhere so there is every reason to believe it occurred here as well. It would be interesting to know the amounts involved. A similar situation would have applied to coal merchants unable to collect supplies from their previous station and now paid to travel further afield.

Overall the opinion of the present writer s that the railway, and government should hang its head in shame.

References at the Wiltshire and Swindon History Centre at Chippenham

2515 Box 11318   117/JS/61/1 13 Barnstaple-Ilfracombe

2515 Box 11334   20/KU.1054 (Part III) 473 Barnstaple-Ilfracombe

2515 Box 11335   20.KU/1054 Part IV 483 Barnstaple-Ilfracombe

36/KU.1055   506
2515 Box 11337   KU.1054/G   541 Barnstaple-Ilfracombe Grant Aid and Closure

2515 Box 11337   KU1/1054 537 Barnstaple-Ilfracombe Financial Information

2515 Box 11343   None 638 Barnstaple-Ilfracombe - Bus Replacement

2515 Box 11352   KU1054 (Part 2) 822 Barnstaple-Ilfracombe

2515 Box 11353   None 825 Barnstaple Junction-Ilfracombe

2515 Box 11354   None 868 Barnstaple-Ilfracombe

Where the Pullman cars of the 'Devon Belle once stood..... . *Jim Gosden*

Issue 14 will see a pictorial continuation of this article with several colour views of the Ilfracombe line in its last years.

Further instalments in the 'To kill a Railway' series are planned, commencing with the Lyme Regis branch in Issue 15.

From the 1938 Southern Railway Magazine

"Driver Silk and Fireman Webb had a shock recently when working the 11.00am 'Atlantic Coast Express' train Waterloo to Salisbury with engine No 860 *Lord Hawke*.

"Just after passing Wimbledon they discovered a large ginger cat crouched in the coal under the left side locker. Although rather dirty, pussy was in good working order and settled down comfortably in the locker for the remainder of the journey.

"At Salisbury the cat was temporarily adopted by the Loco Stores."

As promised from last time, a few electric images, sadly Ken does not appear to have taken many of suburban sets but we hope what follows will make up for that.

Above: Normal working at Shortlands with a 4Cep unit making up a train to the Kent Coast. This is what these units were built for; fast and comfortable services intended to run without delay as indeed they did most of the time.. .

Below: 'Cep to the rescue' No date but Ken was in the right place at the right time (the Southern Region might have preferred him not to have been), when he was able to witness an unidentified Cep set gently buffered up and pushing the stranded 'Golden Arrow' through complete with its failed Class 71 at the head. Even allowing for the blue/grey livery of the boat train this is clearly towards the latter days of the service as out of the 10 passenger vehicles only four are Pullmans. Several heads are observing what is taking place including the conductor at the vestibule of the final Pullman Car. We will never know if the unfortunate passengers missed their sailing. *Ken Wightman / Transport Treasury*

The uncommon pairing of a Class 25, No D5183 and a Class 27 D5381 on the Southern Region. Clearly a freight working we might wonder if this was a failure hence the two locos or simply a means of avoiding a light engine turn.

Something we may be a bit more definite about, a 'Terrier' No 32661 on the Hayling Island branch and form the greenery as well as the number of passengers clearly a summer working. No 32661 was allocated its BR number in May 1951 at which time it had already seen 76 years of service. It would survive a further 12 years but would be one of the lucky to escape into preservation and was instead scrapped in June 1963 five months before the branch itself was closed.

S15 No 30839 running up the South Western main line with the RCTS / LCGB special working of 18 October 1964, 'The Midhurst Belle'. This engine took the first part of the tour from Waterloo after which, 'Six Bells Junction' reports USA No 30064 took over to Guildford and on the Stammerham Junction via Baynards. After visiting Midhurst and Kemp Town the return was from Brighton and the Quarry Line to Victoria behind No 35007.

No 30862 *Lord Collingwood* (one nameplate from which was resented to S C (Collingwood) Townroe at the time of the engine's withdrawal but seen here in service at Basingstoke with a Bournemouth line working. The train is running on the slow line, the fast being cleared for another service; note the starting signal allowing No 30862 to proceed has already returned to 'On' in consequence of the train occupying the next track circuit.

In pristine condition and a cold winter's morning, No 34088 *213 Squadron* has charge of the 'Golden Arrow' near to Shortlands. Compared with the earlier view this time the train is of more Pullmans although with the requisite 4-wheel luggage at the head and possible another at the rear. No 34088 was a Stewarts Lane engine from 1951 through to 1961 after which it moved to the Western section apart from a one month spell at Brighton in June/July 1963. It would cease work from Eastleigh in March 1967 and was cut up by Messrs Cashmore's in South Wales one year later.

Busy times at Bromley South. What could be one of five 'King Arthur' class engines (starting with '307' and ending in '4') is in charge of a Ramsgate working and by the look of it has had some tough times in the past - notice the burn marks at the bottom of the smokebox door, even so there appears to be no shortage of steam and it will also be noted the crew have or are attempting to put on the right hand injector. Alongside is D1 No 31749 the fireman of which is taking the opportunity to bring some coal forward whilst the driver waits for the road. Without an engine number we cannot relate the life of the actual King Arthur class engine but for the D1 further life was limited. Redundant after Phase 2 of the Kent Coast electrification it ceased work in November 1961 and was scrapped.

'Britannia' on the Kent Coast. Whilst most of Ken's slides come with some detail this one has none. For the sake of completeness can someone please help with a location? What we can state is the obvious; named train and early 1950s - *more please!*

The original Godalming station
(with thanks to Colin Martin)

Like many market towns, Godalming had achieved a certain prosperity before the railway came but with its arrival the town flourished. On 15 October 1849, an extension of the Guildford Junction Railway (from Woking to Guildford) was opened to all traffic terminating at a new station to the north of Godalming. On the same day the South Eastern Railway opened its own line from Redhill to Guildford and connecting into the Godalming extension at Shalford Junction.

The residents of Godalming now had direct access to Guildford and London whilst the SER route could take them via a slightly roundabout route to Reading and beyond. As had occurred elsewhere when a railway arrived the previous means of transport, notably the road coaches and in the case of Godalming a Navigation quickly became unprofitable.

The original Godalming station facilities included accommodation for passengers, goods and coal,

wagon turntables allowed for access to the various sidings - some of these remained in situ until as late as 1935. There was also a single road engine shed.

The terminus at Godalming lasted as the principal station in the town for just a decade as on 1 January 1859 Mr Brassey's speculative line from Godalming to Portsmouth came into use and with a through service from Waterloo to Portsmouth.

The route of the new line bypassed the old station instead running to the east and then turning due south. A connection was made to the old station at what was known as Godalming Junction with a new Godalming station, the one that survives to today, almost one mile south of the junction. This possessed only limited goods facilities and these ceased completed with electrification of the main line in 1937. Even after the new line and station were open, for a time passenger services continued to run between Waterloo and what

The facade of the original 1849 Godalming station in June 1937 some years after it had ceased passenger use and was instead the railway goods depot for the town. In outline it was similar to Micheldever although here a number of add-hoc extensions have been made over the years. The net curtains on the upstairs floor windows would imply this was staff living accommodation.
Frank E Box

UP. Week Days.

	1 Empties Alton to London		2 S.E. Goods		3 Goods		4 Passenger to London		5 S.E. Passenger		6 Passenger		7 Passenger		8 S.E. Goods		9 Passenger to London	
	arr.	dep.	arr.	dep.	arr.	dep.	arr.	dep.	arr.	dep.	arr.	dep.	arr.	dep.	arr.	dep.	arr.	dep.
	a.m.	a.m.	a.m.	a.m.	a.m.	a.m.	a.m.	a.m.	a.m.	a.m.	a.m.	a.m.	a.m.	a.m.	a.m.	a.m.	a.m.	a.m.
GODALM. (New)	..	..	..	..	..	..	..	..	..	..	..	..	..	..	..	..	9 12	9 8
Shalford Junction	..	..	..	..	..	..	..	..	..	..	..	..	..	..	..	..	9 17	..
Guildford ... arr.	..	..	..	..	..	..	7 0	..	..	..	8 30	..	8 50	..	..	..	9 12	..
Godalm.(Old) dep.	..	..	5 0	5 0	..	..	..	7 6	8 19	8 19	8 36	..	8 55	..	9 0	9 0	..	..
Shalford Junc. ..	..	..	5 6	5 12	..	..	..	7 10	8 25	8 27	..	8 40	..	9 0	..	..	9 17	9 18
Guildford	2 45	3 0	..	..	5 45	..	..	7 23	..	..	..	8 53	8 58	9 11	9 14	..	9 28	9 30
WOKING	3 25	3 45	..	..	6 0	..	..	7 25	..	..	..	..	..	..	..	..	..	..

	10 Passenger		11 S.E.	12 S.E.	13 S.E. Goods		14 Passenger to London		15 Passenger to London		16 Passenger from Prtsmth.		17 S.E.	18 Passenger		19 Passenger from Prtsmth.	20 Passenger to London	
	arr.	de.	arr.	dep.	arr.	dep.	arr.	dep.	p.m.	p.m.	arr.	dep.	arr.	arr.	dep	p.m.	arr.	dep.
GODALM. (New)	..	..	..	..	..	12 14	..	..	..	1 31	..	..	..	..	..	4 26	..	..
Shalford Junction	..	..	..	..	12 18	..	..	..	1 35	..	..	..	..	4 30	..	..	..	..
Guildford ... arr.	..	..	..	..	12 23	..	..	..	1 40	..	..	..	..	4 35	..	..	..	4 15
Godalm.(Old) dep.	..	10 15	..	..	..	..	1 10	..	..	..	..	..	..	..	..	..	..	..
Shalford Junction	10 21	..	11 0	11	9 12 13	..	1 16	..	..	..	1 24	3 39	3 39	..	..	4 21	..	..
Guildford......	..	10 25	11 5	11 14	12 17	..	12 24	1 20	1 21	1 43	1 30	3 45	3 47	..	..	4 36	..	4 41
WOKING	10 37	10 44	..	..	12 35	1 35	..	1 53	..	..	..	..	..	..	..	4 46	4 53	5 8

	21 Empt. Ash to London		22 S.E. Passenger		23 Empty Fm. Waggons		24 S.E.	25 S.E.	26 Passenger		27 Passenger		28 Passenger		29 Goods		30	
	arr.	dep.	arr.	dep.	arr.	dep.	dep.	dep.	arr.	dep.	p.m.	p.m.	p.m.	p.m.	arr.	dep.	arr	dep.
	p.m.	p.m.	p.m.	p.m.	p.m.	p.m.	p.m.		p.m.	p.m.	p.m.	p.m.	p.m.	p.m.	p.m.	p.m.	p.m.	p.m.
GODALM. (New)	..	..	..	..	..	..	..	..	..	..	..	8 36	..	..	9 23	..	..	..
Shalford Junction	..	..	..	..	..	..	..	..	..	8 40	..	..	..	..	9 28	..	..	..
Guildford ...arr.	..	..	..	..	..	6 40	..	..	..	8 44	..	..	9 0	..	9 35	..	..	10 15
Godalm.(Old) dep.	..	..	..	..	..	..	..	..	8 20	..	..	..	..	..	..	..	..	..
Shalford Junction	..	..	6 13	6 13	6 46	..	7 15	8 5	8 26	..	..	9 6	..	..	10 21	..	..	..
Guildford..	5 35	5 50	6 18	6 19	6 52	7 5	7 20	8 10	8 30	8 43	..	8 45	..	9 10	9 40	10 25	10 45	..
WOKING	6 10	6 55	..	..	7 30	7 35	..	..	8 52	8 55	8 54	8 55	9 25	9 30	9 55	11 5	11 10	..

UP SUNDAYS.

	1 Goods to London		2 S.E.	3 Passenger to London		4 S.E.	5 Passenger to London		6	7 Passenger to London		8 Sheep & Gds. to London		9 S.E. Passenger		10 Passenger to London	
	arr.	dep.	dep.	arr.	dep.	dep.	arr.	dep.		p.m.	p.m.	p.m.	dep.	arr.	dep.	arr.	dep.
	a.m.	a.m.	p.m.	a.m.	a.m.	a.m.	a.m.	a.m.		p.m.	p.m.	p.m.		p.m.		p.m.	p.m.
GODALM. (New)	..	..	..	8 48	..	8 44	..	..	..	..	..	5 30	..	..	..	7 48	7 44
Shalford Junction	..	..	..	..	..	..	..	..	..	..	..	..	..	..	..	..	..
Guildford.... arr.	..	..	..	8 53	..	..	..	..	..	..	..	..	..	..	..	7 53	..
Godalm. (Old) dep.	..	..	..	..	8 35	..	11 35	..	..	5 5	..	5 45	..	..	..	..	..
Shalford Junction	..	..	8 22	8 41	..	10 22	11 41	..	..	5 11	..	..	5 51	7 17	7 17	..	..
Guildford	2 45	3 0	9 27	..	8 54	10 27	..	11 45	..	..	5 15	..	6 0	7 22	7 23	..	7 54
WOKING	3 25	3 45	..	9 4	9 5	..	12 0	12 3	..	5 28	5 30	..	6 30	..	..	6 8	8 10

N.B.—A Special Cattle Train will leave Nine Elms at 12.15 a.m. on Tuesdays and Fridays for Ash.—Empties returning after service.
A Special Cattle Train may be expected to leave Guildford at 3.30 p.m. on 10th and 24th December for Weybridge, Chertsey and Kingston.

---

24   WEEK DAYS.   WOKING, GUILDFORD AND GODALMING BRANCH.

DOWN. Week Days.

	1 Goods from London		2 Gds. from London		3 Gds. from Woking		4 S.E. Passenger		5 Passenger	6 Passenger frm Londn.		7 S.E. Passenger	8 Passenger		9 S.E. Passenger	
	arr.	dep.	arr.	dep.	arr.	dep.	arr.	dep.	arr. dep.	arr.	dep.	arr. dep.	arr.	dep.	arr.	dep.
	a.m.	a.m.	a.m.	a.m.	a.m.	a.m.	a.m.	a.m.	a.m.	a.m.	a.m.	a.m.	a.m.	a.m.	a.m.	a.m.
WOKING	3 50	4 0	4 30	4 35	6 0	6 10	..	7 30	7 38	..	8 16	..	..	8 40	..	..
Guildford	4 15	5 0	5 0	6 30	6 25	6 45	..	7 45	7 50	8 26	8 28	8 30	8 30	..	9 15	10 53 10 54
Shalford Junc...	5 5	..	6 34	..	6 49	..	..	7 48	7 54	..	8 31	8 34	8 34	9 19	..	10 57 10 57
Godalming (Old)	..	..	6 50	..	7 5	..	..	8 0	..	..	..	..	9 25	..	10 57	..
GODLM. (New)	..	5 15	..	..	..	..	..	..	..	..	8 41	..	..	..	..	..

	10 Passenger frm Londn.		11 Passenger frm Londn.		12 S.E.	13 Coals. London to Toghm.		14 Passenger frm Londn.		15 S.E Goods.	16 Passgr. fm Lndn		17 S.E. Passggr.	18 p.m. S.E.	19 Passggrs. fm. Lndn	20 Coals fm. London.	
	arr.	dep.	arr.	dep.	dep.	arr.	dep.	arr.	dep.	arr. dep.	arr.	dep.	arr. dep.	arr. dep.	arr. dep.	arr.	dep.
	a.m.	a.m.	p.m.		p.m.	p.m.	p.m.	p.m.	p.m.	p.m.	p.m.	p.m.	p.m.	p.m.		p.m.	p.m.
WOKING	10 50	11 6	..	..	12 9	..	1 5	1 10	1 38	1 55	..	3 21	3 23	..	..	4 24	5 15 5 20
Guildford	..	11 20	12 19	12 20	12 53	1 30	..	..	2 10	..	2 30	..	3 36	3 30 3 31	4 11	4 34	5 35 6 10
Shalford Junc...	11 24	..	..	12 28	12 56	..	..	2 14	..	2 34 2 34	3 40	..	3 34 3 34	4 14 4 37	..	6 15	..
Godalming (Old)	11 30	..	..	..	12 30	..	..	2 19	..	..	3 46	..	..	..	6 30	..	..
GODLM. (New)	..	..	..	..	..	..	..	..	..	..	..	..	..	..	4 44	..	..

	21 Passgrs. fm. Wkg.		22 Passenger from Lond		23 Coals. London to Alton.	24 S.E. Passenger.		25 Passenger fr. London.		26 Passenger fr. London.		27 S.E. Goods.		28 Passenger		29 S.E. Goods.	
	p.m.	p.m.	arr.	dep.	dep.	arr.	dep.	arr.	dep.	arr.	dep.	arr.	dep.	arr.	dep.	arr.	dep.
	p.m.	p.m.	p.m.		p.m.	p.m.	p.m.	p.m.	p.m.	p.m.		p.m.	p.m.	p.m.	p.m.	p.m.	p.m.
WOKING	..	5 30	..	..	5 50	..	..	7 7	7 9	..	7 45	..	..	9 30	..	..	..
Guildford	5 40	6 5	6 0	6 1	..	6 45	7 17	7 18	..	7 19	7 55	7 57	8 23	8 30	..	9 45	11 18 11 28
Shalford Junc.	6 8	..	..	6 5	..	..	7 22	7 22	7 23	..	8 0	..	8 35	8 35	9 49	..	11 32 11 32
Godalming (Old)	6 15	..	..	..	..	..	..	7 29	..	..	..	..	9 55	..	..	..	..
GDM. (New) dep.	..	..	..	6 10	..	..	..	..	..	..	..	..	..	..	..	..	..

SUNDAYS.   WOKING, GUILDFORD AND GODALMING BRANCH.

	1 Goods fr. London.		2 Goods fr London.		3 Passengr S.E.		4 Passenger		5 Passenger fr. Lond.	6 Passengr		7 S.E.	8 Passengr S.E.		9 S.E. Goods.		10 Passengr		11 S.E. Goods.	
	arr	dep	arr	dep	arr	dep	arr	dep		arr	dep	dep	arr	d.p.	arr	dep	arr	dep	arr	dep
	a.m.	a.m	a.m.	a.m	a.m.	a.m	a.m.	a.m	p.m.	p.m.	p.m.		p.m.		p.m.	p.m.	p.m.		p.m.	p.m.
WOKING	4 45	4 50	5 30	5 35	..	..	11 6	11 9	5 15	5 17	6 25	..	..	..	..	9 28	9 30	..	..	..
Guildford	5 10	7 30	5 50	6 15	9 16	9 17	11 23	11 24	3 30	6 34	6 35	23	23 28	23 24	8 18	8 48	..	9 45	11 18 11 28
Shalford Junc.	7 45	..	6 19	..	9 20	9 20	11 28	..	3 34	..	6 39	7	8 28	8 28	8 52	8 52	9 49	..	11 32 11 32
Godalming (Old)	8 0	..	..	..	11 34	..	3 40	..	..	..	6 52	..	..	..	9 55	..	..	..	..
GDM. (New) dep.	..	..	..	6 25	..	..	11 24	..	..	..	..	..	..	..	..	..	..	..	..

N.B.—A Special Cattle Train will leave Nine Elms at 12.15 a.m. on Tuesdays and Fridays for Ash, and which is expected to arrive there at 2.30 a.m.
An Empty Train leaves Ash Station, (South Eastern Company) for Redhill, every Monday, Wednesday and Saturday morning.

From an original bound copy of the LSWR working time table for January - December 1869. The book contains separate timetables for each month of the year; this is the December 1869 entry. Note also the pages in the original are not bound completely square.

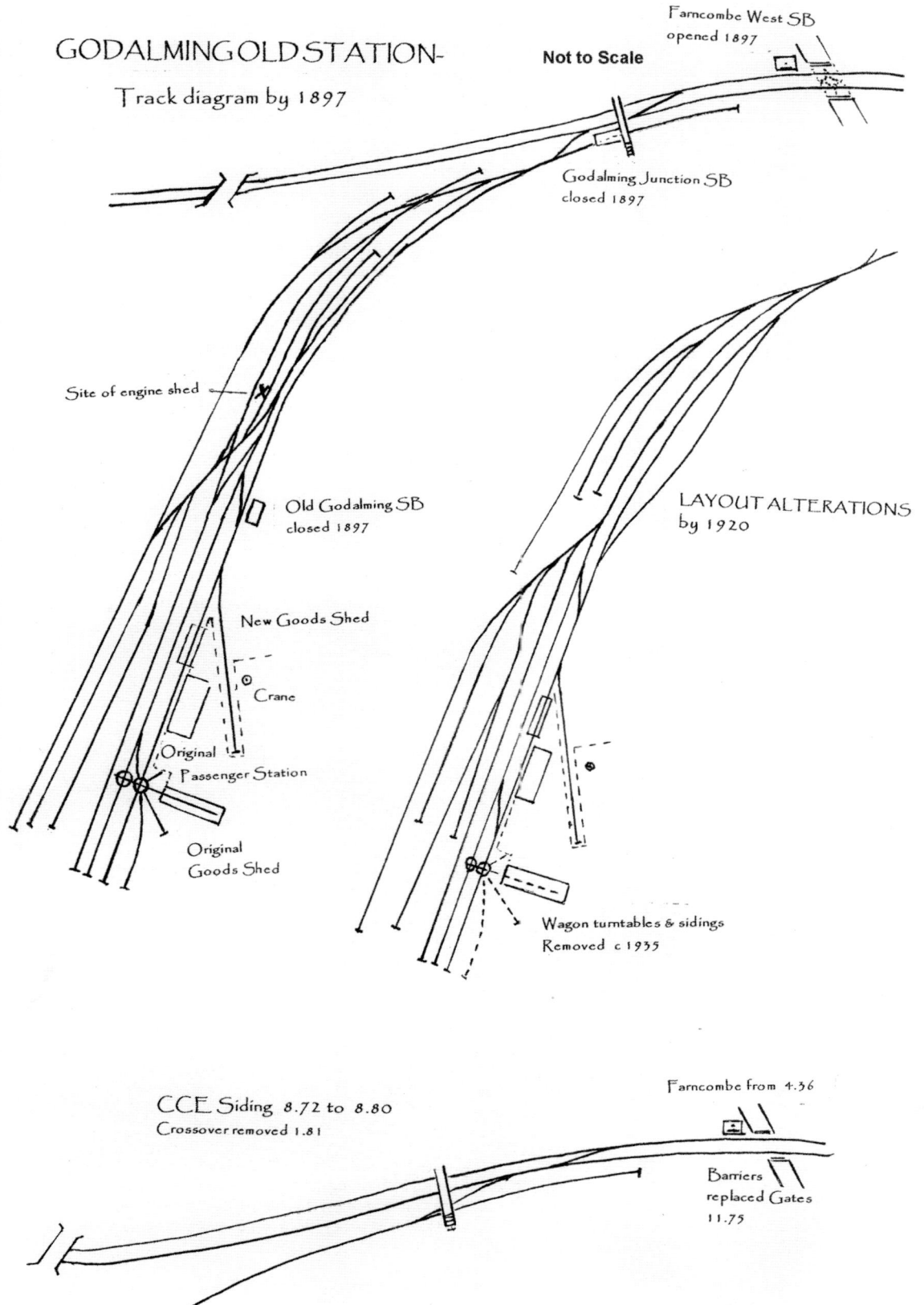

GODALMING OLD STATION-

Track diagram by 1897

Not to Scale

Farncombe West SB
opened 1897

Godalming Junction SB
closed 1897

Site of engine shed

Old Godalming SB
closed 1897

LAYOUT ALTERATIONS
by 1920

New Goods Shed

Crane

Original
Passenger Station

Original
Goods Shed

Wagon turntables & sidings
Removed c 1935

CCE Siding 8.72 to 8.80
Crossover removed 1.81

Farncombe from 4.36

Barriers
replaced Gates
11.75

had become known as Godalming Old, the second station taking the name 'New'. Whether these suffixes appeared at the actual stations or were only in the timetable is not known. Passenger services continued to run to the Old station as late as 1897 at which date a new station at Farncombe on the direct line had been opened and just about half a mile from the original terminus, this then took the traffic for the northern part of Godalming.

All goods traffic was now concentrated on the Old station so much so that an additional goods shed was erected and a number of alterations made to the track layout. The engine was also closed. The site was served by regular stopping goods trains, in the 1950s this was four trains each day, the ability to sideline such

services away from the main line an advantage to the electric service.

Goods continued to be dealt with at the Old station until as late as January 1969, motor lorries having gradually taken the traffic away from rail. In July 1970 the track in the yard was removed but two years later a single siding was restored for use of the Chief Civil engineer when working in the area. This too was removed in August 1980 and plain track laid across what had once been the junction. Since that time redevelopment has meant almost all traces of the old station have been removed whilst only a select few travelling on the main line will be able to recognise what had once been a junction. Locally a new road has also been built into the site with the appropriate name 'Old Station Way'.

Above: Trackside view of the old station showing the goods shed extension and canopy. It is not known if the original passenger station had a canopy on the platform side; note too the varying platform heights the lowest being from earliest days. An 1885 photograph held in Godalming Museum shows 14 staff on the platform c1885. Moving forward 60 years and the Southern Railway had a Godalming and Farncombe Home Guard section. *Frank E Box*

Opposite top: From the junction looking towards the terminus - buildings just visible in the extreme distance. The engine shed had stood between the second and third sidings from the left. *A E Bennett / Transport Treasury*

Opposite bottom: Complete and intact yet devoid of traffic, the scene after railway had acceded to road.

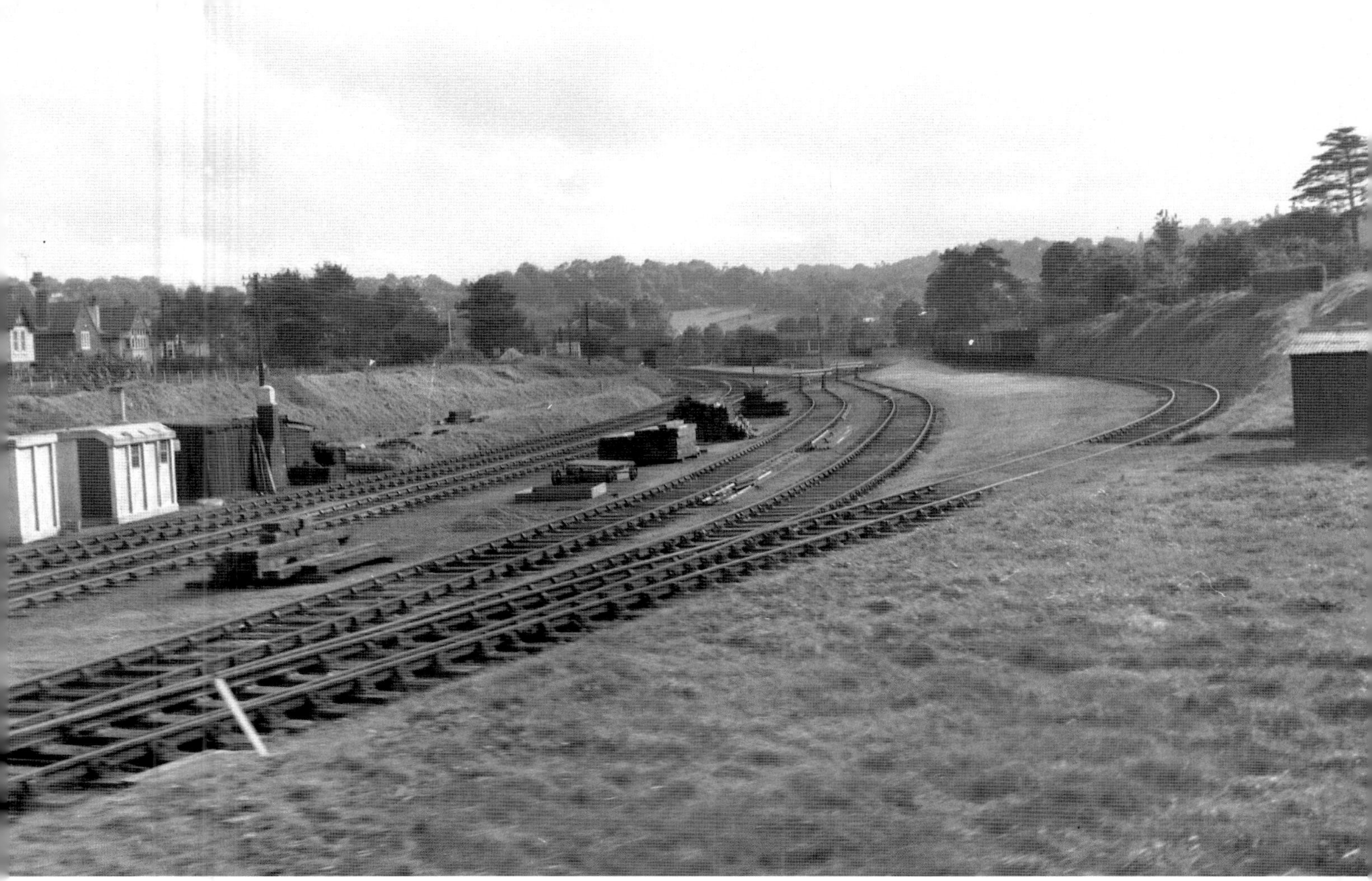

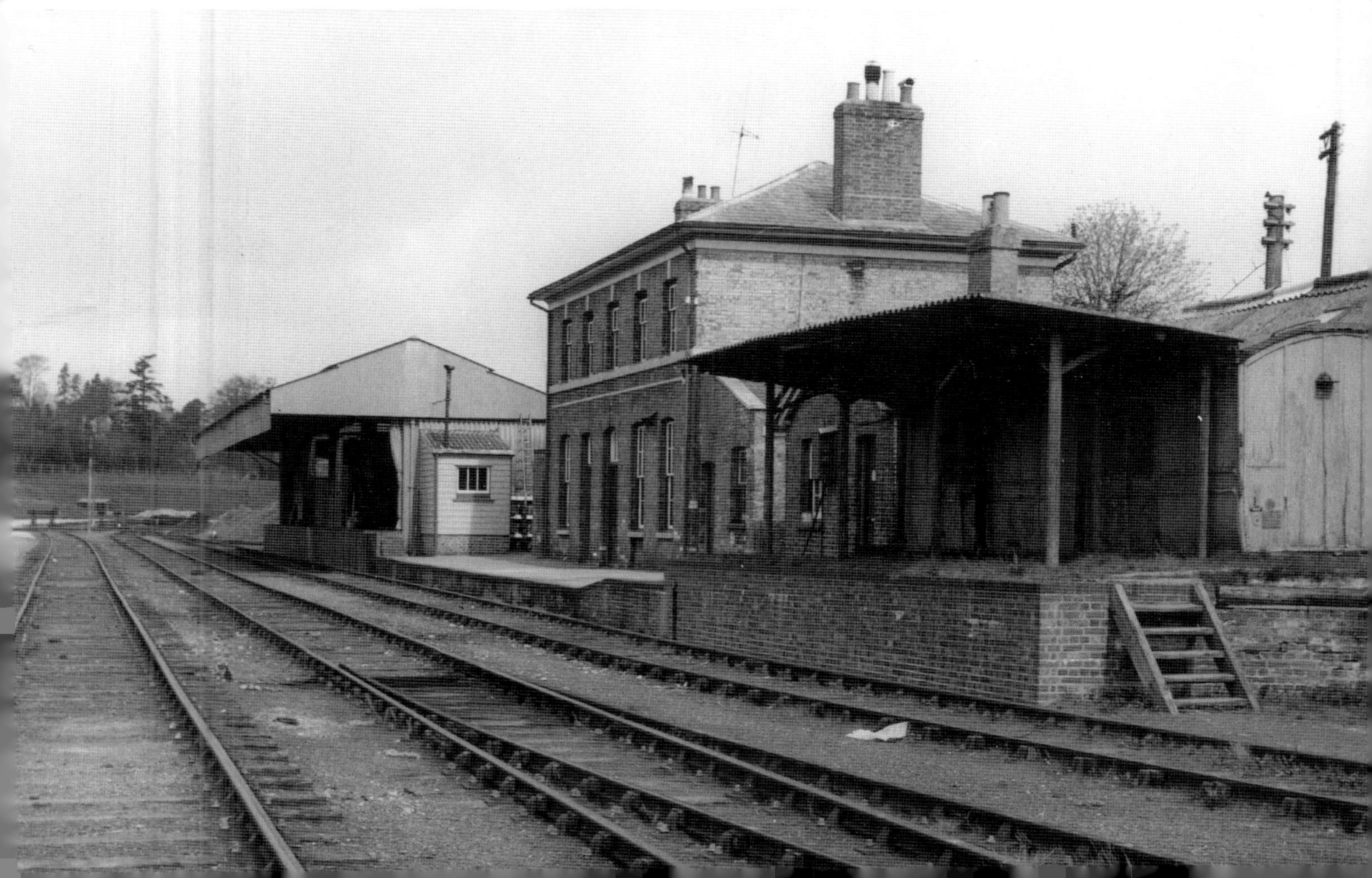

On 25 January 1959 a private tour was run to celebrate the centenary of the Portsmouth Direct line. Just two locos were involved, No 30587 from Victoria to Guildford and then '700' class No 30350 from Guildford to Godalming returning to the main line and thence to Fareham, Gosport, Guildford and back to Victoria. The train was of seven coaches including two Pullman cars. A E Bennett / Transport Treasury

Quieter times with No 31627 on the local goods. Slightly unusual was the position of the loading gauge immediately a the end of the canopy.

The Hugh Davies colour files

Most students of the Southern Railway will know the name Hugh Davies and his 'Photos from the Fifties' business the nucleus coming from images taken by members of the Railway Enthusiast's Club at Farnborough. Hugh would attend railway fairs and events around the Hampshire and Surrey area providing prints of literally hundreds of locations as well as printed lists organised on a county basis. How long he was active in this field is unknown, certainly he was also a member of other organisations including the Stephenson Locomotive Society arranging and also giving talks to members and guests.

Hugh was the most amenable of men, your editor dealt with him on many occasions and it was sad to learn that he was ill and then passed in 2024. His legacy however continues, his website and index still available with the collection now in the hands of the Transport Treasury. The vast majority of the available collection was black and white, Hugh chose not to deal in colour - although he did kindly point the author in the direction of at two colour collections and which have been a most useful source of material. It was a surprise that when the archive was moved from Godalming to High Wycombe to find a number of colour views none of which appear to have been listed but also with no annotation. We are delighted to present a selection possibly taken by REC members. Again if readers can provide further information it would be appreciated.

The Kingsferry bridge on the Sheerness branch provide the backdrop to this view of 2Hap unit No 5607 on a Sheerness service. Notice too the multi-lingual electrification warning notices.

Sidmouth Junction to Exmouth service arriving at Exmouth behind Ivatt Class 2 No 41316. The lower quadrant signals are a delight but sadly are no more now the service is reduced to a basic railway only from Exeter - but at least it survives!

A branch that is no more is the stub of the former Basingstoke & Alton line forming the short section from Butts Junction (Alton) to Treloars Platform. This was a calling point of a number of special workings over the years including this one organised by the Railways Enthusiasts Club. Similar special workings would visit the northern stub at Thornycroft's siding at Basingstoke

Another branch line that survived for goods for many years was that from Botley to Bishops Waltham. Opened to passengers in 1863 the last regular passenger service operated on 31 December 1932 after which it had a life for goods of 30 years as a goods only line. Here parcels are being unloaded from the 'R.U' (restricted use) goods van; seemingly not much other traffic that day. This line also saw several special workings over the years whilst it was also used to test new DEMU units from Eastleigh. The stub end at Botley survives as a headshunt for stone trains.

Southern Railway type 'glasshouse' or 'Odeon' type signal box at deal and at the time of writing still extant. Opened in 1939 it controls the level crossing at Western Road but otherwise the only non main work is to a siding that sees occasional use with Network Rail infrastructure workings.

Instead of 'before and after', this is other way around with the 'after rebuilding' view above, No 35001 at Southampton Central with the 'Up' 'Bournemouth Bells'. Below the nameplate off the engine as affixed to the original design. The sliders for the sandboxes are open whilst the flag, even if a bit faded, is that of the Southern Railway shipping services. Your editor recalls seeing the smokebox numberplate off this engine in the Eastleigh sheds stores in 1964 and made an offer to purchase - number-plates were then going for £1 each. Unfortunately it was already spoken for.

Super-power at Exeter St David's and close to Exeter Middle (also known as 'Red Cow Crossing'). It might be 'Western' but does anyone know the origins of the name? These and their sister tank engines of the 'W' class were ideal bankers most passenger trains having one of these at the rear whilst the Meldon stone trains could see up to three; one at the head and two pushing. Somehow Exeter is not quite the same today.

Strong & Co. were Romsey brewers with a number of tied houses in the Hampshire area. At their peak they owner almost one thousands pubs and advertised their products on the lineside at a number of locations including Micheldever and north of Eastleigh. The artwork was also changed over the years so that in their final form it was an impression of a green rebuilt Bulleid pacific.

Three recent additions to the Bluebell Museum collection:

D. Drummond Esq cot plaque

Staff and others raised money for the London & South Western Railway (LSWR) Servants' Orphanage in various ways. One method of providing funds was for the children's bed, known as cots, to be endowed. Endowing a cot would normally be by a group of workers who would raise the £100 needed. Having endowed a cot the group or person's name would be fixed above the cot.

The plaque on display is on behalf of Dugald Drummond who in 1898 was Locomotive Engineer of the LSWR and was a constant supporter of the Home. His personal contributions amounted to over £1,000, more than £150,000 in today's money.

Thanks go to Woking Homes for loaning the Museum this item.

Dugald Drummond had many rolls prior to becoming Locomotive Engineer for the LSWR in 1895. Born in Scotland he had various jobs there, spent time in Australia before a spell with the London Brighton & South Coast Railway in the 1870's. He died in 1912, aged 72, still working for the LSWR. For the LSWR he designed 22 types of locomotive, one of the most famous being the 'T9', 66 being built between 1898 and 1901. Number 120 is preserved as part of the National Collection.

T9 number 120, in LSWR livery, hauling the Sussex Coast Limited Rail Tour on Saturday 24 June 1962 at Christ's Hospital.
Photo by Joe Kent from the Bluebell Railway Museum, Southern Railways Archive.

A BENCRO Photograph of LBSCR 326 *Bessborough.*

The photograph shows the 1912 built, London Brighton & South Coast Railway (LBSCR) Class J2 locomotive 326 Bessborough pulling the Sussex Belle. The photograph was probably commissioned by LBSCR and taken at the Brighton Works. The backdrop is somewhere else. The photograph was taken in 1912 and used a very special technique of printing. It is in fact two photographs, one printed on the inside of the glass and the other on the back board.

The image takes on a three-dimensional appearance with a remarkable sense of speed despite the locomotive being static when photographed. The technique was known as BENCRO which is an acronym of Benn & Cronin. John Benn was a Southwick and Brighton photographers. They specialised in contract photography for advertisements and magazines and with the LBSCR published the book "The Heart of Sussex".

The J1 and J2 class locomotives

Marsh designed the J1 class following his successful I3 class. They were designed to pull heavier London-Brighton express services. Following Marsh's illness Billington modified the design to the J2 class. Only one of each class was built, J1 No 325 *Abergavenny* and J2 No 326 *Bessborough*. Bessborough hauled the London-Brighton express trains until 1925 when 'King Arthurs' and 'Rivers' replaced them. It then hauled heavier Eastbourne to London express trains until WW2. Both engines were scrapped in 1951.

In the illustration, SR numbered 2326 takes the 11:08am Victoria to Eastbourne via Edenbridge Town and Heathfield on Saturday 10 June 1950 at Polegate. The coaches are one of the many SECR Birdcage Trio-C Sets that operated most secondary services on the Central and Eastern Sections of the SR and BR until mass withdrawal between 1955-1958. *Photo by John J Smith from the Bluebell Railway Museum, Southern Railways Archive.*

Now there are five!

ARDINGLY
FOR ARDINGLY COLLEGE

The nameplate from the Schools class engine 30917.

The Museum recently put on display its biggest enamel sign, one of the Running In Boards from Ardingly station.

The SR target Ardingly with its two neighbouring stations.

With Ardingly appearing twice on this sign the Museum now has five Ardingly's!

The finger board from Haywards Heath.

Next time:

We delve again into the Southern Railway Staff Census.

The Evolution of the Brighton line - Part 1
Jeremy Clarke

It is inevitable that with the passing of time the fifty and a half miles of the former LBSCR main line between London Bridge and Brighton should be looked on as a single entity. But though the majority was produced under London & Brighton Railway auspices, albeit in two parts, it was not built as one, meaning its history is rather more involved than appears at first sight.

There is evidence dating back to the Bronze Age of continual settlement on the coast where Brighton now stands at the point the Wellesbourne river meets the sea. In Domesday the town is noted as 'Brighthelmstone', seeing a growth in size and importance particularly during the Middle Ages before several negative factors caused a decline in population.

Matters started to improve following the London Road being turnpiked in 1807 though with the benefits of sea bathing being promoted, the Georgians had already begun to develop the town as a fashionable resort for the wealthy. It is of no surprise then that the idea of a railway between London and Brighton should be proposed.

William James was the first to put forward the idea, in an essay he published in 1823. It is possible he anticipated locomotive haulage as an engine of Stephenson's colliery type appeared on the cover of the pamphlet. However, it was not simply a London-Brighton line for it also incorporated another joining the two major naval bases at Rochester and Portsmouth: the two routes shared a common portion from north of East Grinstead to a point west of Crawley. The Brighton line would have started at the south end of Waterloo bridge and followed an improved Surrey Iron Railway and its Croydon, Godstone & Merstham extension before a new length to the 'joint' section. The southern part was proposed to use the valley of the River Adur, passing through Bramber to the coast at Shoreham and terminating at Southwick, to the west of Brighton itself. James' ideas proved rather ahead of their time.

Sir John Rennie appears next, stating to a Commons Committee in 1836 that he had been a commissioned to find a line between London and Brighton in 1825 during employment with the Surrey, Sussex, Wilts & Somerset Railway. He selected two routes, one of which was much like today's line, the other heading west from Nine

Steam on the Brighton line, 9 April 1962. No 34008 is at London Bridge with the 4.10 pm working. *Larry Fulwood / Transport Treasury*

LONDON BRIGHTON & SOUTH COAST RAILWAY.

CAUTION TO PASSENGERS

AS TO ALIGHTING FROM TRAINS.—Passengers are cautioned to use great care in alighting from the Carriages; to see that the Train is at the platform, that it is the proper side to alight, and *not to alight till the Train has stopped.*

AS TO LOOKING OUT OF CARRIAGE WINDOWS.—Passengers are also cautioned against putting their heads out of the Carriage Windows when Trains are in motion, and especially *when passing through Tunnels, Bridges, &c.*

THROWING EMPTY BOTTLES OUT OF TRAINS.—Passengers are earnestly requested to abstain from this most objectionable practice, which is attended with much risk and danger to the public, and also to the Company's Staff at Stations, and to the men at work upon the Line.

LIGHTED MATCHES.—Passengers are requested not to carelessly throw away matches in a lighted state, either on the floors of Carriages or out of Carriage Windows, or upon the Platform or Floors of Stations, &c.

COMPARTMENTS FOR SMOKING.—Smoking is only permitted in those compartments of Carriages so designated.

CUSHIONS AND SEATS OF CARRIAGES.—Passengers are requested not to place their feet upon the cushions or seats of the Carriages.

CARRIAGE DOORS & CARRIAGE KEYS.—The locking and unlocking of Carriage Doors should only be done by the Officials of the Company. Passengers are not allowed to use Carriage Keys of their own for this purpose.

London Bridge Terminus. (By Order) **J. P. KNIGHT,** *General Manager.*

Elms and passing through Dorking and Horsham to the coast at Shoreham. Although the SSW&SR failed, revival of a direct line to Brighton resurfaced in 1833. This line was effectively Rennie's, resurveyed under his direction by Francis Giles. Though these plans were also not carried through they came to the fore again in 1835 when proposals for other lines between Capital and coast had appeared.

Charles Vignoles proposed to follow the 'direct line' to Merstham but then head west among the Vale of Holmesdale before turning south for the Shoreham Gap in the South Downs. Nicholas Cundy's line also headed for Shoreham, making use first of the London & Southampton to Wandsworth and then passing through Leatherhead, Dorking and Horsham. Beset with choice, the London & Brighton Committee called on Robert Stephenson to decide which of those offerings was the best. He favoured Cundy's though commenting the line taken was full of errors and, to much dismay and annoyance, promptly drew up and entered a line of his own.

The South Eastern also joined the fray, the engineer being a Mr Palmer: his line through the Downs ran to the east of Rennie's.

Faced with these several routes and the inevitable argument during the hearings, sometimes quite violent, it may be of no surprise that consideration given by the Parliamentary Committee to each proposal was a fairly long drawn out affair. The Committee reported to the House on 13 May 1837 that Rennie's route was favoured. However, the Committee Chairman, Lord George Lennox, stated that on average, attendance of members over the twenty-seven days of hearings had been twenty, but on the day the vote was taken forty-five were there, of whom six voted for Rennie though they had not been present when he gave his evidence. He succeeded in having an amendment passed that an Ordnance Officer be appointed to survey the routes and report on them. Soon afterwards most promoters of the later schemes - Cundy had withdrawn - advised Lennox they had agreed a plan which, together with their own individual ideas, was forwarded to Capt. Robert Alderson of the Royal Engineers. He made his report on 27 June.

While accepting Stephenson's line was to be preferred from the engineering point of view, Alderson recommended Rennie's as the better route. He considered London Bridge a more convenient terminus than Nine Elms, the Greenwich Railway viaduct being easily widened to accommodate the approach. Similarly, he stated Brighton's station was so set as to provide an ideal starting point for a line west to Shoreham and east towards Lewes and the port of Newhaven. Stephenson was outraged, stating that at no time had he been advised a line eastwards was planned. It had not been, Alderson was merely pointed out the possibilty. One change was advised. Rennie had made his route through Cuckfield and Hurstpierpoint to take advantage of the Newtimber Gap in the Downs, but Alderson recommended moving it east to make a more direct approach to the coast. Rennie strongly disagreed and left the project in a huff when the route with this amendment received Royal Assent on 15th July 1837.

As Alderson had perhaps anticipated, the Assent was not simply for a line between Brighton and the junction with the London & Croydon Railway south of Jolly Sailor station – later Norwood Junction – but also for branches to Shoreham and later to Lewes though the latter was nominally independent. Moreover, approval was given for purchase of the Croydon, Merstham & Godstone Railway, dating from 24 July 1805, which the Brighton line followed closely for much of its length.

With the necessary authority and all other matters settled, construction should have started. Unfortunately, someone in Parliament had been looking over a map, coming to a detrimental conclusion so far as the L&BR was concerned. A year or so before the Brighton's Bill was passed the South Eastern Railway had received authority to build a line to Dover: this also made use of the London & Greenwich and the London & Croydon. Originally, it had similarly determined to leave this line at or about Norwood but the SER subsequently settled on using the L&CR only as far as Penge. Thereafter, the route headed rather more toward the south east, climbing the Downs above the Caterham Valley to pierce them beyond Oxted and gain the Eden Valley at Chiddingstone to head for Tonbridge. With the two lines running more or less parallel for the six miles or so between Penge and the point where the SER turned away to the east of Purley the Parliamentarians concluded there would never be enough traffic entering London from the south to justify the arrangement.

Various amendments were thus passed, none of which were happily accepted by either company. The primary amendment ordered abandonment of the SER route between Penge and Chiddingstone and the Brighton line to be used by both companies between Norwood and a point loosely termed as 'north of Earlswood Common' where the South Eastern could conveniently turn away. Cost of construction was to be shared equally. However, a later amendment then stated that rather than the shared portion remain exclusively in Brighton ownership it should be spilt laterally at Coulsdon, the SER taking over the southern half to 'north of Earlswood Common'. This dismayed the Brighton and not simply because the SER had been awarded the more expensive half of the route with its deep chalk cuttings and the tunnel under Merstham for half the overall cost, but mainly because that company now had control of the junction where its line would leave the Brighton one. Moreover, the SER was permitted to purchase the whole of the shared section 'at cost' if it so desired, though never exercised. Despite both companies mounting several challenges to Parliament's basic decision, that and these amendments were neither withdrawn nor changed.

Following Rennie's withdrawal the Brighton company appointed John Urpeth Rastrick as its Resident Engineer. A native of Stourbridge Rastrick was already an engineer of wide experience, working mainly in the Midlands. Among other things, in 1816 he constructed a cast-iron bridge over the Wye at Chepstow and in 1822 was appointed engineer to William James's sixteen-mile long horse-worked Stratford and Moreton Rail Road. In 1808 he built, with John Hazeldine of Bridgnorth, Trevithicks 'Catch me who can' and in 1828, in partnership with William Foster, he had constructed three locos sent to North America one of which, *Stourbridge Lion*, was the first engine to turn a wheel on that continent.

To get to the beginning of its line at Norwood, the L&BR made use of the London & Greenwich Railway. This part of the line had been opened in two sections, between Deptford and a temporary terminus at Spa Road on 8 February 1836 and the mile thence into London Bridge in December that year. (Greenwich was not reached until Christmas Eve 1838.) On then to the London & Croydon Railway which started south from the junction on the Greenwich at Blue Anchor, 1¾ miles out of London Bridge, running for a distance of about 8½ miles: its services began on 5 June 1839, four years to the day from authorisation.

Part 2 of The Evolution of the Brighton line - Part 1 will appear in ST 14.

From the Footplate

Definitely a quieter post-bag this time. Holidays the likely reason or perhaps those day trips are preferable to sitting and typing or running to the post box. Do remember please, we value your feedback, good, mediocre and well 'could do better'. As I have said with previous publications, nothing is perfect but try and accept that we are also human.

To start from **Paul Donnerly** of Hastings. Paul writes some kind words about 'ST' but then goes on to ask a slightly unusual question about resources. 'Dear Kevin. I have followed your writings for many years and also read your comments about ensuring material is safeguarded for future generations. Can I then very cheeky and ask what your own plans are?' I did say it was unusual question but I have corresponded with Paul privately over many years and he is aware I am putting this into this issues.

In short I have made plans already. For personal reasons I have sold most of my 'O' gauge models already. It was a wrench but less for those that follow to deal with. I have also passed on to Transport Treasury two or three 'general' collections of negatives, nothing Southern but items I have accumulated '...it seemed a good idea at the time...' and which will never be used. I do though retain Southern items that could well be useful for article in the future and there are a lot. Eventually though these will be passed to the Bluebell Museum. I would encourage others to think likewise.

Now a note from another Paul, this time **Paul Hepworth** on the pictorial piece on Boat trains and the **Southampton Ocean Liner Terminal - pages 37 to 41 ST12.** Paul asks if it might be possible to have more information on the Ocean Liner special workings

he recalls seeing on the Bournemouth main line in the 1960s? I can answer this one easily as we have been slowly working behind the scenes collecting material on these services for some time and with the intention that it will appear in Issue 14. So, 'watch this space'.

Mike Upton has also been in touch with a couple of points (or should that be 'pints'?) about alcoholic refreshment on trains and at stations. As this will be a subject I am sure dear to many, they are certainly worth including. He starts with some detail on the 'Tavern Cars'. '"...On the bar counter draught Ind Coope and Allsopp's bitter was served straight from the wood, the then new keg beer, 'Double Diamond' as a stable mate. Bottled beers, wines, spirits and minerals were also available."

Going back in time, Mike advises us as to the state of the refreshment rooms on the LSWR back in the 19th century. Waterloo were evidently concerned after complaints by passengers over poor standards in several of the their station refreshment rooms. These facilities were leased to tenants and in June 1882 a contract for all refreshment rooms was given to one firm, Messrs Simmonds and which in turn was taken over Spier & Pond in June 1888. Such was the importance of ensuring alcohol was available to the travelling public that the minute books of the company for 26 April 1882 record, " ...it is essential in the interests of the travelling public that Messrs Bass and Allsopp's beers be supplied at all refreshment rooms "

Evening time at Netley. The Bulleid is possibly on a coast-way working whilst a very grimey Class 3 tank, No 82017 is in the yard. Netley was the junction for the short branch to the military hospital which we hope to feature in later issue.